Great Scientific
Discoveries

Other titles in
Chambers Compact Reference

Catastrophes and Disasters
Crimes and Criminals
50 Years of Rock Music
Great Inventions Through History
Great Modern Inventions
Masters of Jazz
Movie Classics
Musical Masterpieces
Mythology
The Occult
Religious Leaders
Sacred Writings of World Religions
Saints
Space Exploration

To be published in 1993

Great Cities of the World
Movie Stars
Operas
World Folklore

Great Scientific Discoveries

Gerald Messadié

EDINBURGH   NEW YORK   TORONTO

Published 1991 by W & R Chambers Ltd,
43–45 Annandale Street, Edinburgh EH7 4AZ
95 Madison Avenue, New York N.Y. 10016

First published in France as *Les grandes découvertes de la science*
© Bordas, Paris, 1988
© English text edition W & R Chambers 1991

Library of Congress Cataloging-in-Publication Data applied for

ISBN 0550 17002 2

Cover design Blue Peach Design Consultants Ltd
Typeset in England by Butler & Tanner
Printed in England by Clays Ltd, St Ives, plc

Acknowledgements

Translated from the French by Alison Twaddle

Adapted for the English edition by Wendy Lee
Min Lee
Penelope Smith
Brian Williams

Entries provided for the English edition by T J Lee and S Torchinsky, Royal Observatory, Edinburgh:
Aberration of light from fixed stars
Black holes
Cosmic background radiation
Hall effect
Holography
Quasars
Radio waves

Chambers Compact Reference Series Editor   Min Lee

Illustration credits

Contents

Introduction

If we agree that the word 'discovery' implies a find made by chance, and that 'invention' implies a result achieved by knowledge and reasoning, still the demarcation line between the two remains hazy. It is certainly true that it was chance which caused the mould to grow in the Petri dishes in which Alexander Fleming cultured his *Staphylococcus aureus*. The mould destroyed the bacteria, and Fleming concluded that the 'juice' of the mould was responsible for this unexpected effect and christened it penicillin; 15 years or so later medicine was to be completely transformed by this discovery. It is equally true that 20 years earlier a now-forgotten American, M.J. Owens, used his technical skill and reasoning ability to invent a machine which was to have a considerable effect on daily life, since it manufactured bottles which previously had been individually blown. It was, incidentally, bottles produced by the Owens process which were used until 1943 for culturing the miraculous mould called *Penicillium notatum*.

It turns out, however, that Fleming's 'discovery' was not really any such thing. Two and a half thousand years before, the Chinese had discovered that soya cream which had grown mould cured skin infections such as pimples, boils, anthrax and so on; they had in fact used it as an antibiotic salve without knowing anything of antibiotics.

Other specialists before Fleming had also noted the bactericidal power of certain moulds and yet antibiotics as such only came to light after Fleming. Why is that? It is because Fleming, like any inventor, used his scientific knowledge and his powers of reason. That is to say, he understood the import of his discovery and in a certain sense 'invented' penicillin. Pursuing this line of argument, it can also be said that Christopher Columbus 'invented' America. It is generally held that his mission was to find a new sea route to the Indies, and the rather facile assumption is made that his

discovery of the New World was accidental, the result of failure; but it appeared, some decades ago, that Columbus may have been in possession of maps made by a Turkish navigator, Piri Reis, which were a lot more detailed than the European charts. They described both the Americas and the Antarctic. Columbus would certainly not have raised the necessary financial backing for his expedition if he had announced that he was setting out to conquer lands somewhere beyond the Atlantic; maybe he made up the story of a quest for the short cut to the Indies to suit his own ends. Whatever the case he remains, quite legitimately, the 'inventor' of America.

A seed falls on fertile ground

Discoverers are hardly ever ignorant people. An ignorant man may well find a banknote in the street, but if he stumbles upon a hitherto unsuspected physical phenomenon he will not grasp its import. A layman looking through the most powerful telescope in the world, for example, will not recognize an exploding supernova. Steam had certainly lifted the lids of pans since human beings started to cook their food, but it took James Watt, an engineer, to grasp its significance and in the middle of the 18th century he set about researching steam as a motive force. So it can be said that one definition of discovery is the fruit of a seed sown by chance in fertile ground.

Factor one: the state of the sciences

A more in-depth analysis of these conclusions shows that discoveries depend on two social and cultural factors, the first of which is the advancement of knowledge in the various scientific disciplines. So it is that the Chinese, who discovered the bactericidal benefits of certain moulds several centuries before the discovery of penicillin, are not recognized as the discoverers of antibiotics. In fact in their era the existence of germs was unknown, as was the capacity of certain micro-organisms to break them down and destroy them. Even Villemin, creator of the word *antibiotic*, coined after the discovery that certain bacteria attack others, is not credited with the discovery of antibiotics. There was no talk of the discovery of antibiotics until the publication of Pasteur's works on bacterial interactions, notably on the bacteria which attack the *Bacillus anthracis* which causes anthrax; these are the very ones, coincidentally, that Fleming was in the process of culturing when penicillin formed on them. Similarly, in the 3rd century BC an inspired physicist, Hero of Alexandria, also discovered the power of steam. He even constructed the aeolypile, a contraption which worked by steam but was of no real use; his discovery was to lay dormant for centuries because thermodynamics and the dynamics of fluids were as yet unknown.

Factor two: the economic and political climate

The advancement of the sciences is not the only factor affecting discoveries, however. The economic and political necessities of the time are another. When Fleming discovered penicillin in 1928, the possibility of combating germs systematically had not yet been conceived. In his *History of Medicine* Charles Lichtenthaeler recalls a scholar, Ludolf von Krehl, who declared 'when I was teaching in Leipzig around 1890 we did little practical treatment.' He claimed that 'internal medicine was the science of incurable illnesses.' Fleming himself did not grasp that it would be possible to manufacture a drug resulting from his discovery; for a long time penicillin was known only to biologists. Paradoxically the real discoverer of penicillin was the discoverer of sulphonamides, Gerard Domagk, for it was sulphonamides that opened up the era of therapeutics in modern medicine. The imminence of World War II brought a new urgency: enormously high numbers of wounded would run the risk of dying from infected wounds and the pressure was on to

find some way of avoiding that. Chain and Florey took up Fleming's discovery and soon alerted the American government which, motivated partly perhaps by commercial ambitions, mobilized the considerable funding for the purification and mass production of penicillin crystals. It is certain that penicillin would not have appeared on the scene as quickly without the great American laboratories.

Similarly, one might speculate whether the 19th-century expansion of the textile industry, triggered by the advent of the new lower middle classes, was not one of the major factors in another fundamental discovery, that of aniline. At the time the textile industry was a subsidiary of the indigo-dye-importing business which was in the hands of the colonial powers. The Germans, among others, had to buy from the British or Dutch and sometimes from the Chinese or from American dealers who controlled the Central American output. Anyone who managed to break this stranglehold by producing a synthetic form of indigo would pull off a tremendous coup. Might it not have been the hope of doing just that which spurred the young Unverdorben to try to establish the essence of indigo's natural dye? In 1826 he set up a dry distillation on lime. The result was unexpected: an unknown organic base which rapidly went brown and gave off fumes which affected the mind. This was to be aniline. His discovery would continue to have a modifying effect on industrial chemistry because the times were ready for it, indeed were crying out for it. Some years later Runge and Perkin among others explored Unverdorben's discovery further; the formula for indigo itself was not finally established until half a century after the discovery of aniline.

Factor X

It is impossible, however, to ignore the role of chance in the fate of a discovery. Discoveries depend on a number of imponderables, and a major find may go unnoticed for a long time, indeed indefinitely. We will probably never know the worth of 'pancreine', isolated by the Romanian Paulesco and never tested on humans. Perhaps it was another insulin, maybe even insulin itself; perhaps it was the prestige of the American discoverers which unjustly eclipsed the work of the Romanian. It was not that interest in a cure for diabetes was lacking, but Paulesco was unknown and in the field of discovery there is an unconscious tendency to credit only those who already have a good reputation. A discovery marked by a Nobel prize attracts much more attention than the same achievement by an unknown. The 'rich' even share the recognition for discoveries in which they were involved only as patrons; again, the example of insulin bears this out: Macleod, who shared the Nobel prize with Banting, was involved in the work of Banting and his collaborator, Best, only as a patron. In fact he was on holiday at the moment when insulin was discovered! His contribution was in allowing Banting, not without reservations, to work in his laboratory, supporting his work and instructing him in the experimentation processes, a skill sadly lacking in Banting (hence the addition of Best and later Collip). It seems shocking to the contemporary mind to see Banting receive only half the rewards of his discovery, but it is true to say that at the time (and perhaps even today to a certain extent) the prestige of any work is awarded to the head of the laboratory where it is carried out. Lest we rush to write off Macleod as one of science's slave-drivers it is pleasing to note that he was moved to offer half of his remuneration to Collip, the technician whom he felt had been unjustly neglected.

The non-accidental accident

Whatever the role of luck, it must not be over-estimated; we should not suppose that every discovery is an intellectual game where, searching for A one finds B. This is by no means true; a large part must be played by the attitude of the researcher and his observation skills. In the 18th century there was a lot of interest in electricity, which had been known since the 3rd century BC but rather curiously neglected. Whole armies of enquirers experimented with static electricity, the only form known until Volta; armed with pieces of amber, rods of wood, glass, resin or metal, they all rubbed their hardest. A physicist, François du Fay, observed that a very thin piece of gold leaf was attracted by an electrified glass rod, then repelled by another, and then attracted by a resin rod. From this he deduced the existence of two types of static electricity. One he named vitreous, being the property of transparent material; the other he named resinous, presuming it to be the property of resinous bodies. Fay had discovered positive electric charge ('vitreous') and negative electric charge ('resinous'), and had achieved this by experimenting on static electricity and by being a good observer. We may conclude therefore that if a discoverer is someone who, looking for A finds B, he is also someone who, looking for A finds A. Luck there may be, but in scientific research it is not purely accidental.

The role of the printing press

If the growth of knowledge and of scientific and economic (not to mention military) activities favours the proliferation of discoveries — which would explain their huge growth since the Renaissance – it does not follow that no discoveries at all were made in the preceding centuries. The fate of knowledge is closely linked to the spread of printing, and researchers have only been in the habit of recording their work in scientific publications for about 200 years. The wide circulation of publications and the registration of work in indices and data banks have, since then, averted the loss of even the smallest scrap of knowledge. Until this system was set up a discovery was at great risk of being lost, for it was entrusted to only one, or possibly two or three documents, clay tablets, papyri, parchments — fragile objects, to say the least. So we can appreciate what a tremendous loss to us was the destruction of the library at Alexandria. There is good reason to suppose that numerous discoveries perished there. Take, for example, the writing of Erasistratus who, in the 3rd century BC, noted the capillaries uniting the venous and arterial systems. This work would only be rediscovered in 1661 by Malpighi, thereby providing a coherent basis for the concept of the circulation of the blood. There were probably many other discoveries recorded in Alexandria in physics, chemistry and astronomy which were not to be resurrected until centuries later.

Without attributing too much prestige to the knowledge of distant centuries, a fault which has led more than once to ridiculous speculation, it must be conceded that electricity was not totally unknown to the Ancients, as an 'electric battery' once exhibited in the museum of Baghdad and dating from the 3rd century BC demonstrates. Equally we can seriously entertain the idea that the Ancients had some knowledge, accidental of course, of transmutation (the changing of one type of atom into another). It is too much of a coincidence that alchemists chose for their transmutation attempts the two metals, lead and mercury, whose atomic numbers are closest to gold. Atomic numbers, the only logical basis for their efforts, were not established until the 19th century by Mendeleyev; so the alchemists must have had access to some lucky clue which guided their choice. It might have been a natural atomic reactor like the one found at Oklo in Gabon.

Closer to home, there has been more than one discovery which, despite being printed, has been obscured by the negligence or ignorance of those to whom it was addressed. This was the case with a unique article by Gregor Mendel on the laws of heredity, which only survived for posterity by the merest chance. Having appeared in an obscure journal, it was sent in 1866 to several professors but not one took any interest in it. It was 'discovered' 35 years after the monk's death in the archives of the Natural History Society of Brno in present-day Czechoslovakia by researchers who had independently discovered the same laws. What is worse, other works by Mendel, on bees for example, never reached us. They were burned by the monks of his monastery after Mendel's death, because they found the researches of their superior decidedly racy! This touches on the problem of ideology, which will be discussed later.

The domination of Western thought in modern science

It is undeniable that the greater part of scientific and technical advances currently enjoyed by the world community stems from discoveries and inventions made largely since 1750 in the West. In chemistry, physics, biology and knowledge of the physical world, the West has had the lion's share of discoveries. For the East's one Brahma Gupta, studying astronomy in the 7th century, we have Newton, Leverrier, Herschel and Hoyle. For one Ibn Sina, whom we call Avicenna and who, in the 11th century described metals and minerals, we have Davy, Priestley, Lavoisier and Curie. It was as if at the end of the ancient world scientific thought mysteriously went into hibernation, to be reincarnated only for a 'brief' moment between the 10th and 14th centuries by the Arabs. There undoubtedly were intelligent people in the world, yet from China, Japan, and India, the once magnificent and militarily brilliant Orient, next to nothing has come down to us – no exploration of strange lands, no discoveries of physical or chemical phenomena nor even discoveries in medicine, a field full of possibilities for practical application to daily life. It was not that instruments were lacking; any mandarin worth his salt could command a microscope or telescope, retorts or test tubes. No, it seems rather that the main interest was in the power of weapons and that scientific knowledge itself was held to be very much a secondary objective. This fatal error led nations to miss the point that knowledge is itself a weapon. Neither the Indian Mutiny nor the Boxer Rebellion in China would stamp out Western technology. The 'sleeping lands' of the East would only gain their independence by assimilating Western knowledge, if then. The atomic bomb, born out of Otto Hahn's discovery of fission, would overcome the enormous military power of Japan in the 1940s.

The importance of ideology

One could think of many reasons for this great dormant period. One suggestion will suffice. If we examine the two most fertile periods for discoveries, namely the six centuries which straddle the beginning of our era in the Greek world and the period beginning in 17th-century Europe, we conclude that these were periods when successive ideologies collapsed. The Alexandria School, for example, was not subject to any dominant religion; it was a school of criticism. The scientific spirit which began to enquire into natural phenomena in the 17th century, constructing microscopes and telescopes, had itself broken free from obedience to the Vatican. The acquiescence of Galileo, whom the Inquisition forced to deny the Copernican theory, was merely a matter of form. 'Eppur si muove' ('it does move all the same') he is reported to have murmured after his recantation. This procedure was fatal for the temporal authority of the Church, for it demonstrated that faith and more especially the authorities set up to protect it had no scientific competence. Until then there had been no scope for exploring the mysteries of the world around us; all the answers were supposed to be in the Bible. After the Renaissance, however, even this belief was to be questioned by the critical mind and from then on discoveries proliferated.

The danger of scientific dogmas

Not that the forward thrust of science was to be free thereafter. The propensity of the mind to set up as dogma what it knows or thinks it knows supplanted the religious shackles. Ideologies and philosophies appeared and were imposed, holding back scientific progress and discovery for a long time until in their turn they were thrown aside. So the best physicists of the 18th century had a hard job ridding themselves of the woolly theory of phlogistics (a discredited theory of combustion) to which they more or less adhered. Similarly, the 19th-century biologists experienced some difficulty in divesting themselves of the theory of spontaneous generation and of contagion through 'miasmas'.

Some of the greatest minds of the 20th century were disconcerted in the extreme by the propositions of Einsteinian relativity; the great Ernst Mach was to write that he would not subscribe to the idea of atoms and 'other such-like theories'. The eternal inability to entertain new ideas can engender in the greatest scholars, the most perceptive observers, indifference, sarcasm and even reactions astonishingly akin to stupidity. This occurs even in the mid-20th century in the world's most liberal countries and among the scientific community which, in principle, should be the most open-minded in the world.

The most remarkable case of a discovery almost disappearing in this way is that of jumping genes, identified by Barbara McClintock, an American researcher with no fancy titles or financial backing. Her idea, establishing that genes were not evenly distributed and did not follow a line of descent in a rigorously determined, not to say determinist fashion, furiously annoyed American geneticists of the 1930s and 1940s. As for verifying the work of this modest, slightly dotty woman, that would have been too long and painstaking a task. Had it not been for her exceptional persistance, McClintock would have spent 40 years pursuing work which earned her only the derision of her more famous peers. Not until the early 1980s was it finally recognized that the old lady had discovered a crucial phenomenon. She received in quick succession two prestigious awards: the Albert and Mary Lasker prize and then the Nobel prize. Several of her adversaries were long dead by this time. So it was too with the discovery of cerebral hormones, endorphins, an idea forcefully rejected by Sir Solly Zuckermann doyen of British scientific researchers and an expert in hormones. Against all odds, two independent researchers, Guillemin and Schally, obstinately continued their work over the years. They found the endorphins and shared the Noble prize. Yet when Guillemin had submitted an article on his discovery to the well-respected journal *Nature*, he received a reply to the effect that this was the product of his fevered imagination.

There have even been cases of scholars exhibiting a remarkable degree of myopia with regard to their own discoveries. When, for example, the distinguished Otto Hahn analysed the results of an experiment that he had just completed, his mind rebelled. To split an atom of uranium and obtain barium and krypton — such a claim would be worthy of an alchemist. Hahn had in fact succeeded in producing atomic fission (Irène Joliot-Curie and Enrico Fermi had achieved this earlier without really understanding what they had done), but he dared not believe it. The tone of his account bears this out: the physicist does not rule out the possibility of error. It would be his old collaborator Lise Meitner and her son-in-law Otto Fritsch in Sweden who would understand just what Hahn had accomplished and grasp its significance. Four years earlier, in 1935, two scientists of the first rank, Rutherford and Broglie, had smiled when interviewed about the possibility of exploiting atomic energy. 'There is no more reason to think that we will one day construct an atomic-powered engine than to suppose that we will one day construct a conscience-powered engine on the pretext that conscience is the power behind human actions,' Broglie declared to Jacques Bergier. So to avow that we will never again see, as under Charles X, a doctor or zoologist strongly assert that the giraffe cannot exist for reasons of cardiac morphology would be to lay oneself open to the fate of the apostle Peter.

Every discovery is incomplete

It remains to define a discovery. The illustrious epistemologist Karl Popper has devoted a whole volume and more besides to the subject. We shall wisely attempt to give only a brief account of it here, which can be summarized as follows: it is impossible to announce a discovery in definitive terms. A discovery only ever reflects a part of reality.

When they discovered natural mutations, Dubinine and Spenser could not guess that one day the phenomenon observed by one and explained by the other would be found to be far more complex than they had supposed. The discovery of 'useless' segments of DNA, introns, led their followers to think that mutations were not

at all exceptional phenomena and that they occurred constantly, even in the higher animals.

Whether it is a question of a law or an object, a discovery is only ever the acquisition of some piece of knowledge which necessitates a reorganization of previous knowledge. In the case of a law, only repeated verification can allow its elevation to the rank of a discovery. That is the only way to avert a disaster as celebrated as the 'discovery' of the famous N rays by the unfortunate Prosper Blondlot, which was even the subject of a paper delivered to the Academy of Sciences. In the case of an object, only after a systematic analysis can it qualify as a discovery; again this is the only way of avoiding another fiasco like that of Piltdown Man, a fake fossil man made from a hotch-potch of collected bones.

Criteria for our choice

So what are the criteria governing the choice of subjects for a book like this? Why include the discovery of Troy by Schliemann and not that of Tutankhamun's tomb by Carter and Carnarvon? The former revealed that this bed-rock of ancient legends indeed constituted a historical substratum and gave rise to an entirely separate discipline, ethno-archaeology; the latter only yielded, apart from the actual treasure trove, minor information about a princeling of the 18th dynasty and the life and work of ordinary Egyptians of the time. Why include the discovery of the 'Lucy' skeleton, an Australopithecus and not the upsilon particle with a mass six times that of the proton (discovered in 1976)? 'Lucy' has taught palaeontologists a lot about the origin of the human race, while discoveries of elementary particles are so numerous that they could fill a book this size on their own without greatly advancing the general principles of physics.

Our choice has also been guided by the following principle: the most important discoveries are those which affect, not only pure science, but also what might be called our collective knowledge and indeed everyday life to the same extent. The discovery of anaesthesia, for example, has enabled

6

everyone to face up to the prospect of an operation without the terrible distress which prevailed before; the discovery of contraceptive drugs has sparked off a quiet social revolution of which there have been few equivalents in the history of mankind. The discoveries of Neanderthal and Cro-Magnon Man have swept away the myths solemnly adhered to by 'creationists' and at the same time have opened up horizons whose extent we can only dream of, since it is yet possible that the human race as we know it may evolve further in thousands or tens of thousands of years from now. The discoveries of liquid crystals and electromagnetic effects on tumours (Priore's machine) seem to be of major importance, even if incomplete, and that is why they feature in this work along with the discoveries of coffee and rubber.

A further objective of this work has been to inspire modesty by outlining the often chaotic path of knowledge, littered as it is with lost opportunities. What suffering and death would have been spared us if only attention had been paid to the 'antibiotic soya cream' of the Chinese and the bacteria described by Leeuwenhoek. What precious information would have been available to us if, centuries before Niepce, someone had thought of placing a sheet of copper coated with Judean bitumen at the bottom of a black box with a hole in it! Perhaps we might have had a photograph of Julius Caesar or Charlemagne. From the schoolboy to the professor, the mind all too easily falls for the illusion of its own omnipotence.

That our choice has been subjective is proved by the number of discoveries included: some 120 out of what must be thousands. Yet we hope that these limitations will prove useful, for we know from the inventors of the transistor in 1948 that it was the very impurities in the silicon which made possible the passage of the electric current. ...

What happens to discoveries?

The time which often elapses between a discovery and its absorption into the general body of knowledge, that is to say its practical and scientific use, may be so great that there has to be a rediscovery to bring the first discovery back to life. An example of this phenomenon is that of bacteria, which were observed in 1675 by Leeuwenhoek, but forgotten and then rediscovered by Duvaine in 1850. Most often the correct interpretation of the discovery is made by the new discoverer to whom history then accords the credit. One example of this is the case of X-rays, discovered in 1858 by Plucker but rediscovered and correctly interpreted by Röntgen in 1895. The scientific and practical uses of a discovery are often subject to quite lengthy delays. Penicillin, for example, was discovered by Fleming in 1928 but only went into production around 1943.

Key
- ● correct interpretation of discovery
- ○ incorrect interpretation
- ▲ application
- ■ rediscovery

Anaesthesia

Arterial tension

Atomic fission

Bacteria

Benzene

Electric battery

Fertilization

Genetic mutations

Heredity

Ionosphere

Liquid crystals

Penicillin

Photons (simultane

Polymerization

Protozoa

Radioconduction

Sperm

Television

Viscose

X-rays

1772 Davy	▲ **1842** Long		
1733 Hales	▲ **1876** Von Basch		
		1935 Fermi	○ **1939** ■ Hahn
■**673** Leeuwenhoek	● **1850** ■ Davaine	● **1873** ■ Obermeier	
1825 Faraday		● **1931** Pauling	
▲ **1800** ■ Volta			
	● **1875** ■ Hertwig	■ **1882** Fleming	
	1937 Dubinine	● **1957** Spenser	
	1865 Mendel	■ **1900** De Vries Correns Ischermak	
	1901 Marconi	● **1925** Appleton Barnett	
1888 Reinitzer		▲ v. **1960**	
	■ **1928** Fleming	▲ **1943** Florey Chain	
of) **1803** Young			■ **1975** Aspect
	1880	● **1922** ■ Standinger	
1675 Leeuwenhoek	● **1839** Schwann ■ and Schleider		
1835 Munk	● **1888** ■ Hertz		
1677 Leeuwenhoek	■ **1841–49** Kolliker Wagner Leucart	**1875** Hertwig	
1873 May	▲ **1923** Nipkow		
1855 Audemars	▲ **1902** Müller		
1858 Plücker	● **1895** ■ Röntgen		

Discoveries from A-Z

Aberration of light from fixed stars

Hooke, 1669; Bradley, 1725

Capricious stars

*To the earthbound observer the stars
seem to move slightly, but this is
only an illusion.*

At the beginning of the 17th century the Copernican theory of a heliocentric universe — that is, the rotation of the Earth and the planets around the Sun — had been intellectually accepted, but a method of demonstrating it was still being sought. There was a great surge of curiosity about the universe. People wondered whether it would be possible to measure the distance to nearby stars by measuring the angles of the triangle formed by a star and two points at which the Earth is to be found at given times in the course of its orbital journey. (A nearby star will appear to move against the background of distant stars as the Earth moves round the Sun; this apparent movement is called the parallax.) It was not known at that time that such measurements would require extremely precise equipment which was not yet available. The greatest parallax ever observed is smaller than 1 second of arc, but when the 17th-century astronomers set about the task they arrived at astonishingly large parallaxes, which seemed quite plausible to them. The Englishman Robert Hooke observed the star Gamma Draconis in 1669 and concluded that its parallax was 30 seconds of arc. His fellow Englishman John Flamsteed observed the Pole Star for eight years and concluded that its parallax was 40 seconds of arc.

In the 18th century intuition as much as the lack of any coherent pattern in these large parallaxes led the English astronomer James Bradley (1693–1762) to check the measurements of the parallaxes in question.

Assisted by his colleague, Samuel Molyneux, Bradley installed in 1725 a vertical telescope directed at the zenith (directly overhead). It was 7.52 metres long with an aperture of 9.32 cm and was fixed to the chimney of Molyneux's house at Kew near London.

Bradley's observations

Bradley observed Gamma Draconis on 3, 5, 11, and 12 December 1725, and saw no change in the star's apparent position. On 17 December, however, he saw to his surprise that it was transitting more southerly, in a direction opposite to that which would be given by parallax. By March 1726 this displacement had reached a maximum 20

A very difficult measurement
The establishment of a parallax is still a very complex operation today, dependent on numerous parameters. It was only in 1838 that Bessel in Koenigsberg and Struve in Dorpat established simultaneously the first significant parallaxes. Studying Vega with a wire micrometer Struve established a parallax from 0.12″ to 0.126″, the currently accepted parallax being of 0.12″. This information can be used to deduce that Vega is about 26 light years, or 250 million million km away. Bessel, studying 61 Cygni (in the constellation Cygnus) with a special heliometer, fixed upon a value of 0.30″, which is still held today.

seconds of arc. The star seemed to stay in one place for a few days, after which it returned slowly to its original zenith distance. From June to September it moved to the north, and by December was in the same position it had been in during December 1725. Bradley could not explain these changes, either by atmospheric refraction or by parallax. He decided to observe the motions of a large number of stars, to see if they all behaved in this way.

A new telescope with a larger angular range, erected at Wanstead in 1727, made it possible for Bradley to observe nearly 200 stars. It became clear to him that the path of every star observed seemed to vary according to the star's latitude. The fact that the cycle was annual led him to think it must be connected to the earth's motion. One day he took a trip on the Thames and noticed that the flag on the boat's masthead changed direction each time the boat turned, although the wind direction had not changed. He deduced that this effect of relative velocity could equally occur with respect to the Earth's motion and the velocity of light. Bradley found that the aberration of light accounted for the observed star shifts.

This discovery was of tremendous importance to positional astronomy. The observations provided further evidence of

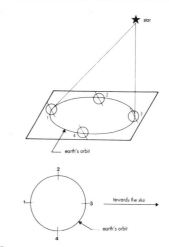

The stars are not where we see them
The movement of the Earth in its orbit makes the time taken for light to reach the observer's eye different at different points of the orbit, except when the Earth is at points 2 and 4, with the star at the summit of an imaginary cone.

the orbital movement of the Earth, and also showed that the stars are very far away. All early star catalogues had to be corrected for the aberrational constant.

Another discovery — nutation

Bradley detected changes in the declination (celestial co-ordinate) which could not be due to aberration. He found variations in the declination of Gamma Draconis and 35 Camelopardis which had a period of about 19 years. Bradley correlated this with the inclination of the moon's orbit to the plane of the Earth's equator, since the line joining the Moon's nodes describes a complete circle in about 19 years. Grounds for the periodical 'nodding' of the Earth's axis, or nutation, had been stated by Newton 60 years before. Now Bradley had observed it.

The absolute speed of light

The discovery of aberration was to lead to the famous experiment of Michelson and Morley completed in 1887. These physicists demonstrated that light had a constant speed in vacuum. Michelson and Morley did not in fact have access to the necessary mathematical tools to explain the phenomenon (the Fitzgerald – Lorentz contraction). It was Einstein who would receive the credit for establishing the speed of light as the maximum speed in the physical world.

Allergy and anaphylaxis

Richet and Portier, 1902; von Pirquet, 1906

The body defends itself

It was not until quite late in the development of medicine that the reaction of the body to invasion by foreign substances was defined.

At the beginning of the 20th century, serums, essential elements in the preparation of vaccines, were the object of systematic research. Two French doctors, Charles Richet and Pierre Portier, were studying the effects of various animal serums. In 1902 they discovered the following: when a dog was injected with a dose of serum from an eel it initially showed no reaction at all. However, the injection of a further weaker dose 20 days later triggered effects which proved fatal. Richet and Portier called this phenomenon anaphylaxis.

The two doctors postulated that the substance responsible for the reaction, the anaphylactoxin, must be poisonous to bring about this pathological reaction. According to their theory the first injection suppressed certain defence mechanisms in the organism, rendering it vulnerable to a second. Subsequent studies indicate, however, that this explanation is false. In fact the precise opposite takes place: the organism does not react to the first injection because it lacks the necessary defences: antibodies specific to the anaphylactoxin. However, from the moment that this invader enters the organism, specialized cells in the immune defence system 'learn' to identify the aggressor. So, with the second injection, mass production of the antibodies is stimulated: the conflict between antibody and antigen (the name formerly given to anaphylactoxins) produces reactions on the surface of certain cells, causing various chemical substances to be produced in the blood, such as his-

tamine, serotonin and bradykinin. These substances alter the whole physiological balance and cause anaphylactic shock. The classic symptoms of this reaction are itching, flushing, a dramatic fall in blood pressure, and breathing difficulties; unless adrenalin is injected without delay, death results. Despite their errors in interpretation, Richet and Portier have the great distinction of having pioneered one of the essential disciplines of medicine: immunology.

Definitions

As early as 1903, the French physiologist Arthus made a major contribution to the understanding of organic reactions to foreign substances: the anaphylactic reaction is not linked to the toxicity of the injected substance. In 1906 the German, Clement von Pirquet, discovered and studied the skin's reaction to tuberculin and added to our terminology the word allergy. This defined anaphylactic reactions to non-toxic substances. In 1910 the Americans John Auer and Paul Lewis established that

---

Dust which makes you ill

The allergenic power of the household dust lurking in carpets and curtains was first identified in 1921. It is explained by the fact that harboured within this dust are dormant bacteria in the form of spores. The allergic nature of hay fever, caused by pollen, and its connection with asthma were established as far back as 1906. On the question of allergies it must be pointed out that people may be allergic to all manner of specific substances and not only to those allergenic foodstuffs most often blamed, such as eggs, shellfish or chocolate.

the major damage sustained in anaphylactic shock in animals is due to spasm of the bronchial tubes and suggested that human asthma could be caused by an allergic reaction.

Nevertheless until about 1950 the distinction between anaphylaxis and allergy remained confused. Clinical studies since then have clarified the difference between the two. Allergy is used in a rather limited sense and simply indicates hypersensitivity to certain substances, whether due to previous sensitization or to immediate intolerance. It has been established that in certain types of allergy, (known as 'tissue' or 'local') there are no antibodies in circulation, in contrast to what takes place in the case of other types of allergy and in anaphylaxis. Anaphylaxis is characterized by the production of antibodies directed against specific groups of substances, particularly venoms and poisons (both animal and vegetable), antibiotics and occasionally certain marine toxins. Most significantly, anaphylaxis is rare in humans while allergy is much more frequent. In the 1960s it was established that psychological factors may play a part in allergic reactions, which is not the case with anaphylaxis.

Related developments

Nearly half a century after Richet and Portier's discovery it was revealed that the intuition of the two doctors had been sound even if their theory had not: anaphylaxis is principally caused by toxic substances. Three-quarters of a century later it has also been shown that the dividing line between anaphylaxis and allergy is not as clear-cut as was once supposed. It has come to be accepted that the severity of the reaction is one of the essential criteria for distinguishing between the two: anaphylactic shock is severe and can be fatal if not treated as a matter of urgency, while allergic reaction is only fatal when it causes acute respiratory failure, typically in the case of asthma. Most importantly, anaphylactic shock seems to occur mainly when the allergen enters directly into the bloodstream, as with an injection. In either case, however, the remedy is the same: injection of adrenalin and antihistamine.

Immunology

The discoveries of anaphylaxis and allergy have not only helped in the understanding and treatment of numerous previously baffling diseases but have also increased our knowledge of certain others which it had been assumed were well understood; thus it has become evident that rheumatic diseases have an immunological component, resulting from an earlier infection. At the first occurence the immune defence systems attack and defeat the foreign substance, but often on the second occasion these same defenders, thrown off balance, attack the very tissues, such as joint cartilages, which they are supposed to protect. This is what is known as an auto-immune disease.

Immunology, based on the discoveries of allergy and anaphylaxis has made grafting possible by the selection of donor tissue from the same immunological group HLA (see p. 93). Finally, the perspective of history reveals the discovery of blood groups (see p. 31), made at almost the same time as those of anaphylaxis and allergy and standing in the same tradition of scientific endeavour: the search for those mechanisms by which the body protects itself against the external aggressor.

Ciguatera and anaphylactic shock

During the summer months toxic organisms pollute certain sea areas in the tropics, particularly in the Pacific. They are eaten by fish, which in turn become toxic without any apparent ill effects. Humans who eat these fish for the first time are quite unaffected, but the second time they become violently ill, with swelling of the face, tongue and limbs, severe disorders of the heart, liver, kidneys and blood, high fever and delirium, all of which characterize anaphylactic shock. This syndrome is called ciguatera.

Anaesthesia

Davy, 1772; Long, 1842; Wells, 1844; Morton, 1846; Simpson, 1847

Conquering pain, conquering taboos

Pain relief during surgical intervention is the fruit of one of history's longest chains of discovery.

The avoidance of pain is an innate nervous reflex and helps to explain the search for anaesthetic substances which began with the birth of civilization. The Mesopotamians in their time knew the whole range of available natural drugs: poppy, opium, indian hemp and mandrake.

These drugs were used differently in different regions and civilizations; the Greeks knew about opium, while Pliny the Elder and Dioscorides mentioned only a mixture of powdered marble and vinegar for severe pain. Such a recipe is not as strange as it sounds, as it causes a significant release of carbon dioxide gas; this puts the patient into hypercapnia (excess carbon dioxide in the lungs or blood), which would be heightened by the absorption of wine and lead to reduction of the pain. In succeeding centuries, on the field of battle, alcohol would be administered along with gun powder, also called black powder.

In the 18th century Franz Mesmer turned to hypnotism to anaesthetize his clients but this process, although enjoying a revival in the 20th century, was not widely taken up. In 1799 the Englishman Humphry Davy, a dispensing chemist, accidentally inhaled nitrous oxide (laughing gas), a gas discovered in 1772 by another Englishman, Joseph Priestley. (The preparation of nitrous oxide is very simple, consisting of gently heating ammonium nitrate in a test tube.) Davy experienced a euphoric state. Following this he organized 'hilarious' gatherings for gas inhalation and discovered that prolonged exposure brought in its wake an insensibility to blows. He proposed, in a paper published in 1800, the use of this gas to render patients insensible before a surgical operation. The idea was taken up successively by his fellow countryman Henry Hickman and by the celebrated French surgeon Dominique Larrey, but in vain. It seems that philosophical ideas caused the scientists' efforts to be rebuffed: pain seemed inevitable and all efforts to avoid it scientifically were thought wicked. It was not until 1844 that a dentist, Horace Wells, extracted a tooth under partial anaesthesia by nitrous oxide. The operation ended badly, however; the patient suffered a lot of pain and Wells was forced to abandon this practice.

'Vitriolized wine'

Despite a certain success the nitrous oxide technique did not last, firstly because it led to physical problems and also because something better had been found. In 1806, the German, Frederic Sertürner, discovered morphine and in 1842 the American, Crawford Long, completed the first operation using ether as an anaesthetic.

Although morphine excited the surgeons of the time it seems that it was never used as an anaesthetic in an operation. The preference was for ether, whose properties had

---

Anaesthesia and the royal seal of approval

Even after his first successful use of chloroform as a general anaesthetic, Simpson did not enjoy the success he might have counted on. The forces of morality rose up against the use of anaesthesia in childbirth on the pretext that it contravened the scriptures: 'in pain will you bring forth your children.' It took the decision by Queen Victoria, head of the Anglican Church, to undergo this form of anaesthesia herself to reduce its critics to silence.

been known since 1730. It was the 'vitriolized wine' or 'ethered spirit' described by the German, F.G. Frobenius. No-one used it as an anaesthetic until the American Thomas Morton held a public demonstration in Massachusetts General Hospital in 1846; he induced total anaesthesia in his patient by the application of an ether pad to the face.

Chloroform

Ether is toxic and the following year, after having tested all possible volatile bodies on himself and his assistants, the Scotsman, James Young Simpson, discovered that chloroform, which had been in production since 1831, could be used as a general anaesthetic, a property apparently unknown to its inventors, Soubeiran, Guthrie and Liebig. Simpson must share the credit for his discovery with the Frenchman, Flourens, who also discovered the effects of chloroform but on animals, and with the Englishman, James Snow, who had simultaneously tested ether and chloroform on humans in the same year, 1847. It is Simpson, however, who is most frequently referred to as the discoverer of chloroform and indeed of anaesthesia, and it is he who was made a baronet in 1866, 14 years after Queen Victoria had decided to be anaesthetized by chloroform for the birth of her eighth child.

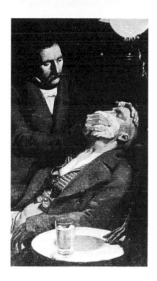

Unsigned painting, 19th century *The first anaesthetics were practised in a thoroughly rudimentary fashion by means of a simple pad of cotton wool impregnated with ether or chloroform and applied to the patient's nose.*

Other methods

The history of anaesthesia is rich in inventions which include: local anaesthesia by cocaine (1884), replaced by novocaine (1904), intravenous anaesthesia after synthesis of barbitone (1902), electric anaesthesia invented by Professor Aimé Limoges (1970) as well as epidural anaesthesia, described for the first time in 1885 by an American neurologist, J. Leonard Corring, and widely used in obstetrics in the second half of the 20th century.

One of the strange things about anaesthesia is that it is not yet properly understood how it occurs. The most advanced neurological researches have not yet uncovered the whole story. We know that in the brain there are specific opiate receptors which pick up the molecules of anaesthetic and whose effect is to reduce sensitivity and depress the central nervous system, but we do not yet know by what mechanism anaesthetics induce a loss of consciousness.

Black powder

The old military recipe which consisted of giving alcohol with gunpowder in order to achieve a degree of insensibility was without doubt a rule-of-thumb procedure but it did have a certain basic merit. In fact black powder contained nitrates. Alcohol, $CH_3 CH_2 OH$, a 'reducing agent' brought into contact with potassium nitrate KNO_3 (the main ingredient in gunpowder), could result in a release of gaseous nitrogen in the stomach, N_2 or nitrous oxide N_2O. The result would be either intoxication similar to the deep-sea drunkenness experienced by divers, or a straightforward depression of the nervous system caused by nitrous oxide.

Aniline

Unverdorben, 1826; Runge, 1834

An eclectic substance

This simple-formula molecule which has considerably enriched the pharmaceutical and chemical industries was in fact discovered in several stages.

In the 19th century, the vegetable dye obtained from two tropical plants of the *Indigofera* and *Ysatis* families was, as it had been for centuries all over the world, a precious commodity. It was used to obtain different shades of blue and especially to dye materials dark or navy blue (otherwise very difficult to achieve). Consumption continued to increase and indigo contributed to the prosperity of the colonial powers. It was easily obtained, first in the form of colourless indican. When dissolved in water the indican easily separated into glucose and indoxyl. By letting this indoxyl dry naturally indigo was obtained. In 1883 Baeyer gave its formula as $C_{16}H_{10}N_2O_2$. Its industrial production began in 1890 and the reign of natural indigo came to an end. Meanwhile another form, much more important, had begun to take over.

Crystalline (aniline)

What Baeyer would do scientifically, a 20-year-old German chemist, Otto Unverdorben (1806–73), had already attempted by craftsmanship half a century before: he had distilled indigo in fairly rudimentary fashion over quicklime. He obtained a colourless substance — hence aniline's first name, crystalline — which had an unpleasant smell and turned brown in daylight or artifical light. Not easily soluble in water, it is an aromatic amine whose formula is simple: $C_6H_5NH_2$. Unverdorben certainly did not grasp immediately the importance of this substance, which struck him as very strange because inhaling its vapours triggered attacks of vertigo and nervousness.

The German chemists took an interest in crystalline: in 1840, Fritsche renamed it aniline, a name derived from the full name of the plant, *Indigofera anil*. In 1834 another German, Runge, broke away from the first material used by Unverdorben and extracted aniline from coal tar. People began to see its future possibilities, at first limited to the textile industry. In 1856 the English chemist, William Henry Perkin, worked on aniline using the inverse process of the one he had followed with indigo: by oxidizing impure aniline he obtained a synthetic dye, mauve — a discovery important enough to earn him a knighthood.

Meanwhile it had been discovered that in the presence of cuprous (copper-bearing) or ferrous (iron-bearing) salts, alkaline chlorates oxidize aniline and produce another colouring agent which has the advantage of being very easy to use: this is aniline black. The history of aniline is punctuated by numerous discoveries, whose authors have remained obscure. One of the names associated with it alongside Unverdorben, Runge and Perkin is that of the Russian, Zinin, who perfected in 1842 one of the three great methods of aniline synthesis: the reduction of nitrobenzene by ammonium sulphide. We have no record of the names of those responsible for the other two processes: catalytic reduction of nitrobenzene, and amination of chlorobenzene by NH_3 at 200°C and under pressure in the presence of copper salts.

A great therapeutic diversity

It was around the beginning of the 20th century that other uses of aniline were discovered. It made possible the synthesis of acetaniline, an analgesic (painkiller) and antipyretic (reduces fever), whose usage would decline with the advent of aspirin. It also made possible the synthesis of atoxyl, which would be valuable in the treatment of syphilis until the arrival of penicillin and which would remain in use against sleeping sickness.

In 1938 Bovet and Halpern, working with aniline, made a completely new discovery. This concerned substances which protect animals against the histamines responsible for allergy and anaphylactic shock (see p. 14): these are derivatives of aniline whose action on blood components had been the object of attention for many years. Since 1942 more than a hundred antihistamines derived from aniline have been launched.

> **Aniline today**
> The two principal contemporary uses for aniline are in the vulcanization of rubber, and in the production of colours for the textile industry, the photographic industry, tanning and the manufacture of paints. Aniline is also used in the manufacture of pharmaceutical products and explosives.

Aniline intoxication

Anilism or chronic intoxication through aniline is an occupational hazard which has been recognized for many years. Its main cause is the transformation of the blood's haemoglobin, whose function is to transport oxygen, into methaemoglobin, which is not suited to the task. One of the first signs of this is a blueness in the extremities (cyanosis); the genito-urinary system is most frequently affected. Primary intoxication by aniline also produces nervous and psychological disorders.

More applications

By 1942 discoveries of new applications for aniline had been going on for nearly a century: certain of its derivatives delay the ageing process of rubber and act as accelerators of vulcanization, the treatment of rubber with sulphur (see p. 221), as discovered by Goodyear in 1838.

While toluidines, anilines derived from toluene, have been used since 1904 in the preparation of TNT or trinitrotoluene (this invention by the German Haussermann in 1891 did not find practical application immediately), aniline was to play yet another part in the chemical and pharmaceutical industries. An excellent solvent of heptane and other hydrocarbons, it was to be widely used in the oil industry and also in the manufacture of two very important groups of products, quinolines and sulphonamides. Quinolines, discovered in fact by Runge in the year that he obtained aniline from coal tar, have only much later risen in pharmaceutical importance as antimicrobial agents. Sulphonamides, discovered in 1938, are also bactericides and were to dominate treatment for infection until the advent of penicillin (see p. 148).

Artificial radioactivity

Curie and Joliot, 1934

Radioactivating matter

The intermediate stage between the discoveries of radioactivity and atomic energy was that of artificial radioactivity.

In the three decades following the discovery of radioactivity (see p. 133) no-one at all supposed that there might exist any radioactive bodies other than those which were produced naturally. However, theoretical physicists took up ideas on the changes of energy in the heart of matter which often went far beyond the field of simple observation; indeed, several of them anticipated important discoveries. So it was that in 1932 many scholars (Anderson, Blackett, Occhialini) discovered positive electrons, put forward as a possibility two years before by the Englishman Dirac. This same physicist, one of the most brilliant theoreticians of quantum mechanics, also postulated that a positive electron, or positron, had only a very brief life and that it destroyed itself on coming up against a negative electron, their simultaneous mutual destruction liberating two photons ejected in opposite directions. The Frenchmen Gibaud and Joliot (the latter being the Curies' son-in-law by his marriage to their daughter Irène) verified Dirac's prediction in 1933. The same year Irène and Frédéric Joliot-Curie repeated the verification with beryllium, which they had bombarded with particles of polonium; they discovered that there was an emission of positive electrons in the beryllium, but that the photons were not emitted. They proposed that the photons materialized in the field of the nucleus, inside the atom itself; this is what they called 'internal materialization', a phenomenon which posed theoretical problems. In fact, having

repeated their experiments with other elements — fluorine, boron, sodium, aluminium — they found that there was also an emission of positive electrons which could not be explained in the normal way. In fact what had happened was transmutation of these elements (see p. 202), sometimes with emission of a high-energy proton, and sometimes emission of a neutron as well as a positive electron.

Radioactive aluminium

What interested the Joliot-Curies in these experiments was the energy balance of the reactions, in the light of which they established that the mass of the neutron was, contrary to what had been thought, greater than that of the proton. Their study of positive electrons, which they called 'electrons of transmutation', had in itself already proved worthwhile by clarifying numerous details of atomic physics.

Isotopes and health

One of the most useful 'spin-offs' of artificial radioactivity was the production of isotopes. These are used, obviously in infinitesimal amounts, as tracers in the study of organic functions; examples are iodine 131, used in the study of thyroid glands, and oxygen 15, used experimentally in the study of respiration.

By continuing this work to find out if neutrons and positive electrons liberated at the time of the bombardment of certain elements were emitted simultaneously and if they occurred for the same energy threshold, the Joliot-Curies discovered another unexpected phenomenon: positive electrons continued to be given off from the bombarded aluminium, even after the bombardment had been stopped. The aluminium had therefore become radioactive in turn.

Neutronic bombardment

The discovery was published in the *Comptes rendus de l'Académie des Sciences*, dated 15 January 1934; obviously it caused a sensation. Scientists from many countries repeated the experiments, with aluminium and also with many other elements. The first bombardments were carried out using deuterons and protons accelerated by a voltage multiplier, by an electrostatic generator and even by a magnetic resonance accelerator or cyclotron. The elements did not all lend themselves equally well to artificial radioactivation, until the day in 1935 when the Italian Fermi established that the Coulomb field of the nucleus was not so strongly opposed to the entry of a neutron as it was to the entry of a charged particle. From then on we turned to bombardment by neutrons to produce artificial radio-elements.

Radio-elements

At the end of 1935 100 artificial radio-elements were known; by the end of 1939, 330; by the end of 1943, 420; in 1957, more than 1000; and in 1970, more than 1500.

An earlier discovery

Although defined in 1934, artificial radioactivity had in fact been discovered along with natural radioactivity by Marie Curie in 1898, when the physicist perceived that exposure to radioactive bodies rendered certain substances radioactive themselves. It might even be maintained that the discovery of artificial radioactivity went back to the use of the first cathode tubes and the production of the first beams of neutrons. In fact, when Röntgen produced X-rays for the first time (see p. 223) he was producing beta radiation comparable to that emitted when matter disintegrates. However, the times were hardly able to grasp the concept of natural radioactivity; as for the artificial variety, it was far too soon for that.

Towards atomic fission

The neutrons used for the manufacture of artificial radioactive elements can be obtained from chain reactions, particle accelerators or natural radio-elements. In fact artificial radioactivity can be produced with any type of particle, provided that they can be accelerated to a high enough speed to penetrate the target nucleus; for example, X-rays, gamma rays, ions of helium, deuterons, protons . . .

Useful in industry for the manufacture of isotopes, the discovery of artificial radioactivity and its corollary, transmutation (see p. 202) has in fact been an indispensible stage between the discovery of natural radioactivity and that of fission (see p. 203). The disturbance of the atom's equilibrium, with the consequent release of energy in the form of an emission of particles, photons, electrons and heat, was to guide physicists unerringly towards the liberation of growing masses of energy under special conditions. What was striking about this process was that the later discovery was both inevitable and accidental.

Astronomical telescope

Anon

A millennial instrument

It seems likely that the invention of the first telescope for observing the sky was totally accidental.

The invention of the telescope is currently attributed to the Dutchman Hans Lippershey, of Middelburg, and the date of this invention is given as 1608. This is a controversial claim, because Lippershey, having manufactured and sold several astronomical telescopes to the Netherlands government, wanted to protect his business by a patent or exclusive right to manufacture for 30 years and this demand was refused; in fact, the government, which was only interested in magnifying telescopes for military reasons, objected that several other people already had knowledge of this invention.

The first telescope

According to other historians, this invention was due to Galileo who, in 1609, was the first to train a telescope on the sky, in order to observe the stars. However, it is an established fact that telescopes were then being sold in Paris, London and numerous towns in Italy and Germany. Galileo simply perfected the commercial telescope of his day, on the basis of optical principles which were in fact rudimentary. The first 'telescope' which he produced, in a single day, consisted of a convex lens and a concave lens fixed at opposite ends of a lead tube; in this way a magnification of three times was obtained. It was following this that he perfected his own adaptation, leading to the construction of an instrument with a focal length of 4.4 cm, which allowed him to achieve an angular magnification of 33 times. He was then able to observe the Moon, the satellites of Jupiter, and the phases of Venus, and to discover that the Milky Way is composed of stars.

In fact it appears that the astronomical telescope was not an invention but a discovery, very probably accidental. Where was this discovery made? Perhaps in Egypt, since glass was being manufactured and polished there in 3500 BC. Equally, glass lenses have been found dating from 2000 BC in Crete and Asia Minor. It is very probable that they were first used as magnifying glasses. (Nero, who was short-sighted, in fact wore a monocle of polished glass.) The fact remains, though, that the knowledge of optics was then too rudimentary for someone to have thought of placing two lenses on an axis in order to obtain a particular magnification. This first telescope was doubtless invented by the accidental placing of two lenses, one on top of the other. So there were in fact astronomical telescopes in existence long before Lippershey.

The proof
Astronomical telescopes existed before Lippershey and Galileo. The evidence for this is that in the 13th century Roger Bacon was writing that with lenses one could read the smallest letters at an incredible distance and make the Sun, Moon and constellations appear much larger than they look to the naked eye.

Atomic fission

Birth of the bomb

The possibility of splitting the atomic nucleus shocked even the most eminent physicists.

After the discovery of artifical radioactivity by Irène and Frédéric Joliot-Curie in 1934 (see p. 20), a great many physicists throughout the world tried to explore the phenomenon with the aim of establishing a working model of the atom and the forces governing it. The principle of artificial radioactivity is established as follows: accelerated particles bombard particles attached to the nucleus of an atom, energy is given off and an isotope of the bombarded atom is obtained.

In 1935 in Rome, Enrico Fermi, discoverer of activation by neutrons, was not however very sure that this was what happened when uranium 238 was bombarded with accelerated particles. With his colleagues Rosetti and Agostino, Fermi bombarded some uranium and obtained four radioactive elements or artificial isotopes whose half-lives were 10 seconds, 40 seconds, 13 minutes, and 1 hour 30 minutes. The last two did not correspond to any known uranium isotope. It is true, though, that at the time work was being done on very small quantities of uranium and that, with its 92 protons (hence its atomic number, 92) and its 140 neutrons, uranium 238 perhaps held some surprises. Fermi concluded that elements existed which were heavier than uranium and could be produced by bombarding uranium with neutrons; these were called transuranic elements, to which he assigned the atomic numbers 93 and 94 and gave the provisional names ansenium and hesperium.

The radioactivation of uranium was also posing problems for the Germans Hahn,

Meitner and Strassmann, who found several other transuranic elements which they numbered 93–7 and to which they gave other names: ekarhenium for ansenium, ekaosmium for hesperium, then ekairidium, ekaplatine and ekaor.

The situation was complicated by the fact that other physicists found more and more transuranics. In France Irène Curie and P. Savitch also tackled the problem and found another transuranic with a half-life of 3 hours 50 minutes; they did not think too much of it, however.

An 'impossible' result

Then in January 1939 news broke of Hahn's publication of an article on a most unusual experiment. In the review *Naturwissenschaften* Hahn declared that he, like others, had bombarded uranium with neutrons and that he had obtained ... barium! Renowned chemist that he was, Hahn couldn't be mistaken, the scientific world

---

A risky experiment

When he bombarded uranium with neutrons in his laboratory at the Kaiser Wilhelm Institute in Berlin (now the Max Planck Institute), Otto Hahn ran the risk of unleashing a real explosive atomic reaction. Fortunately there was only a very small percentage of fissile U 235 in his sample of uranium, and the reaction stopped for lack of 'fuel'.

was sure of that. Now, a barium atom contains 56 protons. If the nucleus of the uranium atom had been split by fission, only 36 protons would remain, that is to say krypton. It was unthinkable that one could split an atom of uranium and obtain barium and krypton. Deeply troubled, Hahn concluded his article in unusual terms: 'as nuclear chemists, in some ways close to physics, we cannot yet bring ourselves to make this leap, contrary to all current experience of nuclear physics: a series of extraordinary coincidences could have led us into making a mistake.'

Hahn's reputation, however, excluded all possibility of error. The whole scientific world understood the significance of his extraordinary experiment. The Dane Niels Bohr, creator of one of the great models of the atom, crossed the Atlantic to inform the Americans of the development and explain it to them. In a few days the Carnegie Institute in Washington, Johns Hopkins University and the University of California set up experiments designed to study the process of fission, barely believable though it seemed; the same thing was happening at the Physics and Technical Institute in Leningrad, in Warsaw and at the Collège de France.

It became clear that the mysterious transuranic elements which had been obtained for some time were in fact true products of fission (hence Fermi's uncertainty); but no-one as yet could see how neutrons had been able to produce fission of the massive uranium atom and throw up both a barium and a krypton nucleus, while unleashing energy of some 200 million electron-volts. (The actual term 'fission' is due to the Austrian Lise Meitner, Hahn's collaborator, exiled in Sweden.) Meanwhile, however, Hahn had succeeded in producing fission of an atom of thorium.

The explanation came from the Collège de France, Joliot, Kowarski and von Halban stated that fission produced in its turn so-called secondary neutrons which maintained the disintegration of the nuclei by a series of other fissions, each neutron liberating two others in a chain reaction. This was exactly what the Hungarian Szilard had postulated in 1935.

The struggle for the bomb

It was Niels Bohr who, in 1939, pointed physicists towards invention of the atomic bomb by suggesting that there were two types of atom in uranium: one with 92 protons and 146 neutrons, U 238; and the other with the same number of protons but only 143 neutrons, U 235. Bombarded by neutrons, the first type of atom, which represents 99.3 per cent of the total, absorbs them with no further consequence; the second becomes subject to a chain reaction.

In September 1939 war broke out. The Germans, who had at their disposal experts of the calibre of Hahn, threw themselves into the problem of atomic energy which each side realized could be used to invent weapons of unprecedented power. Britain cut off Germany's supply line of Congolese uranium, then the only known source, and France seized the only known stocks of heavy water (see p. 88). In September 1939 Einstein and Szilard pressed President Roosevelt to sanction the building of an atomic bomb, but even Einstein, at that time, hardly believed in it. Six years later, however, it was ready.

Scepticism

In 1936 the Englishman Rutherford and the Frenchman de Broglie both expressed their strong scepticism concerning the possibility of exploiting nuclear energy and, in 1939, Einstein had little faith in it. In 1942 the first nuclear reactor was built in Chicago and in 1945 the first experimental atomic explosion took place.

Australopithecus

Dart, 1925

An ill-received ancestor

*The first of these ancestors of man
was discovered thanks to a series of
accidents.*

In 1925 a young South African noticed, while visiting a friend, what she took to be the skull of a baboon. The owner of the object said that it couldn't be because it was at that time widely assumed that anthropoid apes, the ancestors of man, had lived in tropical forests. Since there had been no such forests in South Africa this could not be a baboon, which is classed among the anthropoid apes. The fossil, for such it was, came from the quarries of Taung in Botswana, (formerly Bechuanaland). These quarries were very rich in fossils but until then no-one had paid them any attention.

An off-course baboon

The visitor reported the fact to her anatomy professor, Raymond Dart. He confirmed the absence of tropical forests in South

Skull of australopithecus
*Australopithecus africanus Taungs, found in
South Africa.*

Africa, and indeed of any traces of anthropoid apes which might have inhabited them — none had ever been found in South Africa — but he challenged the thesis that baboons, which are certainly anthropoid apes, can only live in tropical jungles. These animals are also suited to life on the ground in an arid climate. According to Dart they had been present in South Africa for thousands of years.

Nevertheless, intrigued by the baboon's skull fossil, Dart had sent to him two crates of the fossils found in the quarries at Taung. He found, among other remains, an impression of the interior of a skull which intrigued him; then he found the fragment which fitted it and understood right away that this was not a baboon skull.

A human skull

Too many characteristics differentiated this fossil from the large apes. The most important difference was that the foramen magnum, the hole in the skull through which the nerves of the spinal cord pass to reach the brain, was situated at the base of the skull as with humans, and not at the back as with apes. This proved that the creature to whom this skull had belonged had adopted an upright stance a long time before; this would not have been the case with any ape. Furthermore the skull was high and rounded and the arrangement of the teeth was characteristically human, not including the large canines of the chimpanzee or those of the baboon.

The Child of Taung

The face was encrusted with limestone, which Dart spent 73 days scraping off — with makeshift tools, for he knew nothing of the techniques of palaeo-anthropology. (He even used one of his wife's knitting needles.) When he had cleaned it he found himself face to face with the skull of a young child.

At that time the idea of the 'missing link' between ape and man was exciting much interest. Dart sent an article to the prestigious journal *Nature*, which published it. He called the individual which he had discovered *Australopithecus africanus*; the public called it the Child of Taung.

Aged between 3 and 5 years, the child of Taung could not be dated at the time of its discovery, as present-day methods did not then exist. According to latest estimates, however, this descendant of Lucy (see p. 120) would have lived 2 million years ago, 1 million years after Lucy and another million before the appearance of *Homo erectus*.

Inauspicious beginnings

The Child of Taung's debut was difficult; almost everyone in palaeo-anthropology, while not denying the exactness of Dart's description of the skull, refused to make definite acknowledgment of its race. Dart's lack of experience in this area did not do much to help matters. His thesis was supported by one lone specialist, the illustrious Broom, who succeeded in turning the tide of opinion in favour of the discovery and in opposition to Keith, the leading figure of British palaeo-anthropology who then dominated the international scene. Keith, who has been much maligned, had good reason to be hesitant about Dart's discovery: the skulls of Java Man and Heidelberg Man certainly made a case for a transition from the great apes to man, but the evolutionary scheme of things had been set back by the famous Piltdown Man, which had a man's skull and an ape's jaw. It was only in the 1950s that it became evident that this skull was an out-and-out fake (may be made as a means of revenge by Sir Arthur Conan Doyle, the 'father' of Sherlock Holmes, who was exasperated by scientists' hostility towards spiritualism, in which he strongly believed).

In 1938 Broom himself found, also in South Africa, a second skull, related to Dart's but different enough for it to be seen as a descendant of *Australopithicus africanus*; this was *Australopithecus robustus*. In 1947 Keith at last admitted that the theories of his adversary Dart were well founded.

The man of fire

At that time Dart, who had all but been driven out of his mind by the tribulations of the Child of Taung, had discovered another fossil which he named *Australopithecus prometheus* because it was, he said, acquainted with the use of fire. Furthermore, Dart launched himself into rather wild theories about the 'savagery' of man's ancestors. A rather more scientific climate did not begin to establish itself in palaeo-anthropology until the 1960s, when it became evident that the numerous discoveries achieved in this field since the end of the 19th century did not by any means cover the whole range of evolutionary stages between the great apes and the human race.
(See pp. 49 and 120)

No descendants
The Child of Taung belonged to a branch of evolution destined for extinction, since it seems that its descendant, *Australopithecus robustus*, did not survive beyond another half million years, nor did it leave any line of descent.

Barium

Scheele, 1774; Davy, 1808

An avalanche of discoveries

Barium, potassium, sodium, strontium, calcium and magnesium; two chemists discovered this series of simple elements in the space of about 30 years.

In 1774 the Swedish chemist Carl Wilhelm Scheele, discoverer of tungsten and oxygen, found that a mineral called pyrolusite contained an unknown base which he called baryta, from the Greek *barys* ('heavy'), because its weight was considerable; this mineral was in fact barium oxide. He sent his colleague, Johan Gottlieb Gahn, discoverer of manganese, samples of sulphate crystals of the same element which he had produced himself. Shortly afterwards Gahn in turn discovered that a fairly heavy phosphorescent stone, then called Bologna stone, was in fact composed of this same barium sulphate; but neither man identified the simple element within these compounds.

The isolating of barium

The electric battery had just been perfected when in 1808 the English chemist Humphry Davy (barely 20 years old but already head of the laboratory at the Medical Pneumatic Institution in Bristol which specialized in the study of gases) resolved to use it for electrolysis (breaking down a liquid by passing current through it). He was inspired by the work of the Swedes Hisinger and Berzelius, who had shown that the current from a battery decomposed saline solutions, and also from the work of Berzelius and Pontin, who showed that by using mercury as a cathode (negative electrode) it was possible to recover the products of decomposition. This is how Davy first isolated barium alloyed with iron. Contrary to what its name might lead one to suppose, barium is a relatively light metal.

Davy then isolated strontium from strontiarite, potassium from potash in sheets, sodium, calcium and magnesium, and so extended the list of simple elements. Although he applied his methods to other areas — aluminium, glycine and silica, — he did not achieve any concrete results but suspected, rightly, that they did indeed contain metals. Others too would make finds: Saint-Claire Deville in 1854 with aluminium; and Wohler and Bussey with glycinium (now known as beryllium). In the case of silica, a compound from which the metalloid, silicon, was isolated in 1823 by Berzelius, it is possible that Davy did not differentiate precisely between silica and silicates, the latter comprising in combination with silicon, lithium, sodium, potassium, magnesium, calcium, strontium, barium, manganese, iron and aluminium. Be that as it may, Davy made giant strides in chemistry. From then on Davy himself, between 1808 and 1810, Gay-Lussac and Thénard had at their disposal elements which would allow their researches to advance even more quickly.

> **Too short a life**
> Born in 1778 at Penzance, Cornwall, Sir Humphry Davy died in 1829 in Geneva. His contribution to scientific and technical knowledge was immense. Over and above the discoveries described above, he explained the bleaching action of chlorine, proved that diamonds are a form of carbon, invented the miners' lamp, and studied agricultural chemistry, voltaic batteries, the nature of fire, the mechanics of volcanic eruptions and much more.

Benzene

Faraday, 1825

A mere curiosity

The simplest of all the aromatic hydrocarbons was for a long time a mere curiosity of chemistry.

In 1825, an amateur engineer, Michael Faraday, then aged 34, whose scientific interests were divided between chemistry and physics, carried out some experiments in distillation. Among the products on which he worked was whole oil, from which he extracted a gas. Out of curiosity he set it alight and the gas burned.

Benzol
(crude benzene)

For many years the benefits of what Faraday called 'bicarburet of hydrogen' were unknown. In 1833 the German, Eilhardt Mitscherlich, rediscovered the same gas by heating benzoic acid over lime and he named it benzine, a name which would later be reserved for the distilled form of benzol. In 1846 the German, Hoffman, then in 1848 the Englishman, Mansfield, isolated the same product in large quantities by distillation of coal tar; in 1866 the Frenchman Marcellin Bethelot was the first to achieve synthesis of benzene, which was then called benzol, from acetylene. For a long time benzene could only be extracted from coal but with the growth of oil exploi-

Formula of benzene
In this singular structure the atoms of carbon are separated by a distance of 1.39 angstroms, an intermediate measurement between the normal distances in single and double carbon bonds.

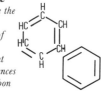

tation it was given a new lease of life. From the start of the 20th century benzene, a colourless volatile liquid with a characteristic odour, saw an extraordinary growth in its application in industrial chemistry by virtue of its properties as a solvent. It dissolves rubber, resins, waxes and though only slightly soluble in water, it is soluble in alcohol, ether and acetone. Benzene does not easily undergo addition reactions but has a large number of substitution reactions, giving a wide range of aromatic compounds.

Faraday probably had no idea when obtaining his 'bicarburet of hydrogen' that his element would long remain an enigma to chemists. In order to explain its properties the German chemist, August Kekule, proposed in 1865 a hexagonal formula with an atom of carbon and an atom of hydrogen at each corner with alternate double and single bonds between carbon atoms, guaranteeing the necessary symmetry to keep account of the number of known isomers. It was only in 1931, and by resorting to quantum mechanics, that Linus Pauling established that the molecule of this chemical, the simplest of all the aromatic hydrocarbons, had a hybrid structure. The normal distance from one atom of carbon to the other in a double bond is 1.34 angstroms and in a single bond 1.50 angstroms but the atoms of carbon in benzene are 1.39 angstroms apart, an intermediate and unusual distance. (An angstrom is equal to 10^{-10} metres.) This fact has been verified by X-ray analysis.

An exceptional career
Born in 1791, Michael Faraday was at first an errand boy in a bookshop, then an apprentice bookbinder. These occupations gave him a taste for books, especially scientific works. He began to study at evening classes and lectures, including some given by the famous Davy. He was to become his assistant in 1816.

Black holes

Collective, 1983

Chasms in the sky

---

*A celestial body whose gravity is so
strong that not even light can escape
its pull.*

There are few concepts stranger to the lay-person than that of 'black holes', massive celestial objects whose escape velocity approaches the speed of light; this means that no light rays, or indeed other electromagnetic radiation, can be reflected or emitted by them. Such an object would appear entirely black, hence the name 'black hole'. The first candidate black holes were celestial bodies located in double star systems: binary systems Cygnus X-1 observed by the X-ray satellite UHURU in 1972, and LMC X-3 and AO620 00 measured by the X-ray satellite Einstein in 1983. When material from the less massive star of the pair enters the gravitational field of its massive but invisible partner it is accelerated and gives off X-rays as it is accelerated or accreted on to the black hole. The mass of the invisible object can be calculated from an analysis of the light (and the periodic variations) from the visible partner. Invisible objects with a mass greater than a certain limit must be black holes. Since these first candidates more such binary systems have been identified. However the most massive black holes are at the centres of galaxies, for example the object NGC 4151 has a black hole of 100 million times the mass of the sun, and it is now commonly believed that the numerous quasars are powered by black holes. Most remarkably there is very strong evidence for the existence of one or more black holes close to the centre of our own Milky Way galaxy.

A 'Dark Star'

The theory of black holes as an astronomical and physical concept and a metaphysical and philosophical curiosity goes back to the 18th century. The French mathematician Pierre Laplace postulated the existence of bodies so massive that light could not escape from them. It is much easier for a spacecraft to lift from the moon than from the Earth. This is because the pull of gravity on the moon is only one sixth of that on the more massive Earth. A rocket needs to be moving much faster when it attempts to leave the Earth's surface than it does when it leaves the moon in order not to be pulled back down to the surface, ie to reach escape velocity. In the same way the pull of gravity from objects many times as massive as the sun would require an escape velocity greater than the speed of light.

Laplace supposed that light was composed of particles as in Newton's composition theory and therefore that such stars would emit no light; they would be invisible.

Eventually Einstein's theory of general relativity, in which the equation of mass and energy is proposed, confirmed that Laplace's ideas were well founded. Black holes cannot emit or reflect electromagnetic photons, that is to say they cannot be observed with optical, infrared, or radio telescopes. To detect candidate black holes, astronomers observe the behaviour of material the most likely explanation for which is that it is in the vicinity of an object with a mass many times that of our sun; if such an object is not shining brightly in the optical or infrared it must be a black hole.

Dying stars and black holes

How might black holes originate? One possibility is that they are the final state of certain stars. A star starts its life as a huge mass composed of gas and dust (but mostly hydrogen) collapsing under gravity. As the star collapses it becomes very hot (several millions of degrees) and the heat energy slows the collapse because the gas atoms attain speeds approximating the escape velocity. Hydrogen nuclei in the star fuse to form helium. This nuclear fusion reaction releases energy because the helium nucleus weighs slightly less than the four hydrogen nuclei which fused to form it. The centre of the star thus generates its own heat and maintains the outward pressure which withstands gravity and prevents further collapse of the star. In time the hydrogen is used up and the core of the star is mainly helium; less heat is produced and the pressure in the centre drops. Gravity again takes over and the star collapses heating up the helium. Eventually the core becomes hot enough for the helium nuclei to fuse together to produce carbon. Now one nucleus is created which is lighter than the three helium nuclei which formed it; again the extra energy is released. When stars are sufficiently massive a succession of phases of collapse, heating and fusion reaction occur to produce heavier and heavier elements until a core of nuclei of iron (the most tightly bound nuclei) has formed.

The final fate of the star depends on its initial mass. Possibilities are a white dwarf, a neutron star, or a black hole. In white dwarfs the pull of gravity is held up by what is called electron degeneracy pressure. For more massive stars, degenerate electron pressure is not enough to counteract its gravitational force, the core of the star collapses violently and at very high densities the electrons and protons combine to form neutrons. If degenerate neutron pressure can provide sufficient force against gravity to prevent collapse a neutron star is formed. When the core collapses it releases a vast amount of energy very suddenly to blow off its outer regions in a violent supernova explosion.

Very massive stars, greater than about 50 times the mass of our sun, undergo a different fate. The core that remains has a mass so great that even degeneracy pressure of neutrons cannot restrain gravitational collapse; the gravitational field becomes so large that no light can escape. Only a gravitational force, a black hole, remains.

Blood groups

Landsteiner, 1900; Jansky, 1907; Moss, 1910

The identity of blood

*It was only advances in microscopy
which permitted the complete
classification of the four major blood
types.*

Ever since the Renaissance audacious doctors had attempted blood transfusions which had all ended in catastrophe, so much so that these attempts had been forbidden. For centuries no-one knew why the blood of one individual could not be injected into another, since the constituents of both seemed identical.

Incompatability of blood

In 1895 the Frenchman Bordet discovered that the red corpuscles of an animal species agglutinated (clumped together) in the presence of serum from another species; this was hetero-agglutination, a prelude to the discovery in 1900 by the Austrian Karl Landsteiner, of the same phenomenon in red corpuscles and serum of the same species. This time it was a case of iso-agglutination. He undertook the experiment again with the blood of his collaborators and made the fundamental discovery of substances named agglutinogens which, when they meet equally specific substances, which he called agglutinins, in another blood type, produce the fatal agglutination. He found the agglutinogens of two types which he called A and B and their corresponding agglutinins, which he called alpha and beta. The presence of these at last explained blood incompatibilities.

Landsteiner established three types of blood: in addition to A and B is type O, which does not contain agglutinogens. In 1907 Jansky, and in 1910 Moss added to these type AB. Thus the four major blood groups we know today were defined, established according to agglutinogens and agglutinins which are in fact the antigens and antibodies specific to blood.

Having emigrated to the United States, Landsteiner discovered, in collaboration with Levine in 1927, two secondary groups, MN and P. Later the Americans Kell, Duffy, Lutheran and Lewis would discover several others. In addition to Landsteiner's classification, which consists of six possible safe combinations of blood groups (group O being considered as universal donor), can be mentioned those of Duffy, Kidd, Cellano and Lutheran which consist of many more combinations (489 888 in Kidd's classification).

> **The compatibility game**
> An A person can only receive A or O blood; a B person only B or O blood; and an O person only O blood. An AB person can receive A, B or O blood. For this reason O is called the 'universal donor' and AB the 'universal receiver'.

Person	Blood		
	A	B	O
A	x		x
B		x	x
O			x
AB	x	x	x

It became possible to achieve blood transfusions without risks, which enormously reduced mortality in operations and mortality due to accidents. The blood bank concept had been born.

Nowadays agglutinogens can be simply removed from blood by 'washing' it with physiological serum. The agglutinogens of A and AB blood types are the most resistant and require 10–20 washings. For the other types 2 or 3 are enough.

The rhesus factor

In 1939–40, Landsteiner, in association with Wiener, completed the fundamental discovery of the blood groups with another just as important find: in 85 per cent of white race subjects, the red corpuscles are agglutinated by the serum of rabbits which has been immunized against the red corpuscles of the *Macacus rhesus* monkey. This indicates that there exists in certain red corpuscles an antigen identical to that of the rhesus, which led to the establishment of a hitherto unknown factor, called simply rhesus. The following year Landsteiner's old colleague Levine demonstrated it was this rhesus factor which explained the haemolytic disease of the newborn, due to an incompatability of the rhesus factor between the blood of the mother (rhesus negative) and that of the foetus (rhesus positive). The blood of the mother in fact destroys the blood of the foetus.

A therapeutic instrument

Another very important chapter in medicine, that of pre-natal therapy, had just begun: thanks to premature delivery by Caesarian section at the 34th week and an exchange transfusion of the newborn's blood (with rhesus negative blood), it is possible to save the baby's life.

The discovery of blood groups, and consequently that of blood antibodies, was also to bestow a precious tool on medicine, namely the tracking down of numerous infectious diseases thanks to the rise in knowledge of specific antibodies. The first application of this method was to be detection of syphilis by the Bordet–Wasserman method in 1906 (now abandoned in favour of more precise techniques).

An instrument of genetics

This discovery was of equal benefit to genetics: it is now known that a person from group A has parents AA, AO or OA; a person from group B parents BB, BO or OB; a person from group O only parents OO; and a person from group AB only parents AB. This law has repercussions for the establishment of paternity.

Landsteiner's discovery even allows for a reconstruction of the migratory journeys and meetings of ethnic groups. It is known in fact that Africans are 42 per cent group B and 99 per cent rhesus positive that Caucasians or white populations 45 per cent group A, 40 per cent group O and 85 per cent rhesus positive, while the Indian population of America are 67 per cent group O and 99 per cent rhesus positive. Further haematological studies allow us to connect these groups with particular sensitivities to certain germs, for example Koch's bacillus for group O type.

Attempts to relate blood groups and psychological characteristics were even to be undertaken several times and some of them indicate that people in group A have a greater capacity for learning than others do; these conclusions remain hypothetical.

> **The genetic game**
> The gene which produces O blood being recessive, it is not expressed in the phenotype, or physical make-up. This is how type A subject can have as parents AA or AO subjects. However, an AO or BO man and an OO woman can give birth to an O subject if the gene transmitted by the father is O.

Blood pressure

Hales, 1733

The detector of arteriosclerosis

The connection between the pressure of arterial blood and that of the atmosphere took nearly a century and a half to attract the attention of the medical world.

In 1733 the English pastor, physiologist and botanist Stephen Hales was taking a blood sample from the carotid artery of a horse, with the help of a fine tube or cannula connected to a narrow glass tube which he was holding vertically; the original aim of the experiment was to establish the speed at which the blood was ejected. Hales was surprised to discover that the blood spurted to a height of nearly three metres. He was the first to establish a connection between blood pressure and atmospheric pressure, or in other words to measure the former.

This fundamental discovery had practically no effect on medicine for the next 150 years or so; blood pressure was then held to be of no interest and it was thought that a vein or artery would have to be incised to measure it, thereby risking a haemorrhage.

Two pressures

It was not until 1876 that Von Basch produced the first apparatus capable of measuring blood pressure without cutting into an artery. It was a complex piece of laboratory equipment and, even after being simplified in 1896 by the Italian Scipione Riva-Rocci, it remained rather impractical. The principle was accepted, however: by putting pressure on the area by means of an inflatable tourniquet to which a manometer is attached, it is possible to measure blood pressure. The descending pressure, the minimal, is that which occurs at the moment when the heart contracts and is called systolic; the rising pressure, the maximal, recognized by a deep vibration caused by the strong pressure, is that which occurs at the moment when the heart relaxes and is called diastolic.

In 1905 the Russian N.S. Korotkoff found the means of measuring diastolic pressure simply by applying a stethoscope at the point where the pulse is taken. Meanwhile the equipment derived from Hales' discovery had been greatly simplified and was widely used. Its merit lay in forging an awareness of hypertension, an ailment not recognized until the early 20th century and the causes of which began to be understood only in the 1950s. Hypertension is the consequence of troubles affecting the internal wall of the arteries, essentially *arteriosclerosis*, of *dysfunction* of the kidneys and the adrenal glands, of an imbalance between certain parts of the nervous system (the *sympathetic* and the *parasympathetic*), and of a problem in the production of a hormone secreted by the heart.

A forerunner of genius

The true worth of Stephen Hales (1677–1761) has not always been recognized. After his experiment on blood pressure, Hales correctly measured the capacity of the left ventricle (chamber) of the heart, the cardiac output per minute, the speed of the blood flow in the vessels and their resistance to it. A talented inventor, he also perfected a system of artificial ventilation which helped to improve the quality of life in ships and in prisons and which was also used to ventilate grain silos.

Carbon dating

Libby, 1946

The memory of an isotope

The intuition of one of the 'fathers'
of the atomic bomb helped to create
one of the most useful tools of
archaeology.

Willard Frank Libby (1908–80) was not only an eminent chemist, but also one of the collaborators in the celebrated American 'Manhattan Project', which from 1941–5 worked towards the invention of the first atomic bomb. His contribution was in fact essential to the separation of the isotopes of uranium, itself vital to the manufacture of the bomb. An isotope of a particular atom has the same number of protons in the nucleus as the original atom but not the same number of neutrons, which explains why the isotope does not have exactly the same chemical properties.

In 1946 Libby demonstrated that in addition, among natural radioactive substances like uranium 235 (called primordial because they have existed since the formation of the Earth), there are some which are formed by the constant bombardment of the Earth by cosmic rays, high-energy particles which have the same effect as radioactivity in a laboratory. The first element formed by cosmic radiation — and therefore called cosmogenic — found by Libby was tritium, the heaviest isotope of hydrogen. Libby soon realized that tritium was certainly not the only cosmogenic isotope made by cosmic rays. He reasoned as follows: when the rays reach the Earth, they begin by crossing the atmosphere, which contains nitrogen. After collisions in the upper atmosphere, these rays contain neutrons, and an atom of nitrogen contains 7 protons and 7 neutrons. If a neutron strikes it, it may eject a proton, which then captures a free electron and thereby becomes an atom of hydrogen. The original nucleus of nitrogen has thus become a radioactive isotope of carbon, carbon 14, with 6 protons and 8 neutrons.

A half-life of 5730 years

Libby reasoned further that this isotope must react very rapidly with the oxygen in the air and produce carbon dioxide (CO_2) and thus the radioactive carbon must be diffused widely in nature; it must therefore also be found in plants and animals, but in

> **Multiple uses**
> Dating by carbon 14 has been applied to tens of thousands of objects, despite its cost. It has exerted a major influence on the archaeology of unknown sites, enabling it to be established, for example, from what date a site was occupied by the comparative study of the ages of different artefacts. In association with stratigraphy (the study of rockstrata), with magnetometry (the study of the magnetic orientations of artefacts such as pottery), and with thermoluminescence it is one of the classic tools for the study of the past. It is equally valuable in the detection of fakes as long as they contain organic particles. (Neither stone nor metal objects lend themselves to the process.)

infinitesimal quantities. Libby estimated at the time that there must be one atom of carbon 14 to a trillion atoms of ordinary carbon; and carbon 14 must have been accumulating in this way since the origins of the Earth.

Libby first tried to measure the atoms of carbon 14 in the environment and to verify that the distribution of this isotope was uniform; he would then be able to examine whether the amount of carbon 14 in ancient objects was the same as in the present day, acknowledging the fact that this isotope has a half-life of 5730 years (at the end of this time interval it has lost half its mass by radioactive decay). So Libby got hold of some Egyptian objects whose date was pretty certain, and which went back some 5000 years. In making compensatory calculations Libby found that the amount of carbon 14 in the environment is indeed constant, at around 5 per cent. So it was that the first measure of dating by radioactive isotopes was established.

How the dating is done

The principle of dating, as it was perfected, is as follows: carbon 14 is only found in living tissues or those which have been alive, because they have maintained exchanges with the atmosphere. So it is only valid for wood, materials, grains, leather, bones and shells, as well as areas containing organic debris, like peat. This method of dating cannot be applied to mineral materials which have not had exchanges with the atmosphere. The measure of an object's age is achieved by the comparison of the specific radioactivity of a current living example and that of some specimen which has been dead for a long time. For example, if in a living tissue X disintegrations per minute are registered and in the specimen $X/2$ are registered, it can be estimated that the carbon 14 of the dead specimen has lost half its mass; therefore it has completed one entire half-life and is about 5730 years old.

Rectifying the calculation

The calculations, in reality more complex, must also take account of the site where the specimen has been found. In volcanic regions, for example, the measurement may well be lower than normal due to the fact of volcanic carbon oxide diluting the carbon 14 in the vegetation. It is also known that since 1900 the relative amount of carbon 14 has been decreased (by 2–3 per cent) because industrial activity has pumped considerable quantities of carbon dioxide into the air; on the other hand, it has increased again by some 50 per cent since the 1950s because of atomic experiments. Some supplementary corrections were incorporated for certain periods of ancient history when it was established by historical comparisons that, around the year 6200 BC, the amount of carbon 14 was 8 per cent higher than today's level; an object whose real age would be about 8500 years would only appear, by carbon 14 dating, to be 7500 years old.

From 500 to 50 000 years

The precision of carbon 14 dating diminishes the further one goes back in time, and it reaches its minimum point at around 50 000 years; one then has to use other methods of dating. At the other end of the scale the minimum point is around 500 years. For objects of this age or less other techniques of identification are used, such as thermoluminescence, notably in the field of art.

Dating by carbon 14 requires particular precautions in the choice and preparation of samples, for the isotope used has only very weak activity and demands very great precision of measurement.

Other methods

For periods earlier than about 50 000 years ago we must turn to other techniques of radio dating, to uranium 238 or 235, to thorium 232, to potassium 40, or again to rubidium 87, but these techniques are essentially used in geochronology (the study of time in relation to the history of the Earth). The choice of technique also depends on the type of geological stratum to be studied.

Cerebral hormones

Guillemin, Schally, 1970

The glands of the brain

*The secretion of hormones by the
brain was a fact that stood neurology
on its head.*

At the beginning of the 1950s, a Frenchman, Henri Guillemin and a Pole, Andrew Schally, who were studying together at Montreal University, were struck by the hypothesis of the Englishman Geoffrey Harris: the pituitary gland, control centre of endocrine function situated in the brain, might be dependent on hormones secreted by the brain itself.

A bold hypothesis

Harris's proposition, which was not supported by a single fact, appeared more than a little absurd; the best neurophysiologists of the day were certain that the brain could not be compared to an endocrine gland, that it could not be a producer of hormones. Sir Solly Zuckermann, a leading figure, went to great lengths to demonstrate the futility of experiments which attempted to explain Harris's theory. However, Guillemin and Schally began to pursue, independently, research into an unknown cerebral hormonal factor which might be secreted, as was very tenuously indicated, by another part of the brain, the hypothalamus. Although bitterly opposed to each other, Guillemin and Schally both supposed that the factor in question would trigger the secretion, by the pituitary, of the hormone which in turn controlled the adrenal glands: ACTH or corticotrophin. In fact, according to their thinking, it was the hypothalamus which governed the 'conductor of the endocrine orchestra', the pituitary. That is the reason why both of them

called the factor they were looking for CRF, corticotrophin releasing factor.

They then had to isolate whatever quantities they could of CRF in the hypothalamus of animals; as the hormone was only to be found there in amounts too small to be discerned by chemical analysis, they needed thousands of samples. It was partly due to the amount of relatively expensive primary material required that these two researchers remained alone in their field. In 1960 Guillemin was working in Paris at the Collège de France and it was fairly easy for him to go round the abattoirs, whereas Schally, working in Houston, had a more difficult task.

3 milligrams

Nevertheless it was Schally who, after working on the hypothalamus of 100 000 pigs, was the first to succeed in gathering 3 mg of a substance which in its chemical properties very closely resembled the CRF he was looking for; he called it TRF, thy-

An exceptional antagonism
Relations between Guillemin and Schally were so poor that, when the Frenchman asked the Pole to send him some samples of TRF discovered by the latter, Schally took refuge behind some hypothetical American law forbidding the transport of chemical substances from one country to another.

rotrophin releasing factor, because it seemed to be concerned with the secretion of thyrotrophin, a thyroid gland hormone. Guillemin himself had as yet found nothing.

Both these men were subsidized by research bodies and, having spent a lot of money buying animal tissue, all in the pursuit of an objective deemed illusory by the best experts, they were each asked to submit their findings to a conference to be held in Tucson, Arizona, in January 1969. Everyone realized that the object of this conference was to attack the 'absurd' theories of an endocrine-type function of the brian. Only Schally seemed to have any worthwhile results to present, but only three weeks before the conference Guillemin had a double success: not only did he discover the TRF already isolated by his rival, but he went further and established its formula. If the conference was only a semi-success, many neurochemists remaining sceptical, it did nevertheless ensure the continuation of subsidies for the two researchers. Their work could go on.

Both men in fact postulated that the hypothalamus actually secreted two hormones, the second being concerned with the genital system; they both called it LRF, luteinizing releasing factor. At the end of 1970, at almost the same time, Guillemin

---

The hormone that couldn't be found

One of the intriguing points of the efforts of Guillemin and Schally is that the hormone for which they were searching in the first place, CRF, has still not been found; it is not even known whether it exists. It could be said that they arrived at their discovery by starting from a false hypothesis.

---

and Schally isolated LRF, and this time it was Schally who announced its function.

It was no longer possible for even the most hostile neurophysiologists to contest the discovery. It was Guillemin particularly who took well-earned revenge on a scientific community which had, in this instance, been singularly lacking in an open-minded and objective spirit. In fact when Guillemin had sent a communication to the American journal, *Science*, the organ of the American Association for the Advancement of Science and a serious publication of international authority, his paper had been rejected, on the surprising pretext that the TRF supposedly discovered by Guillemin was simply the product of his fevered imagination! In 1978 these two scholars shared the Nobel prize for medicine and were, for the first time, obliged to shake hands.

The role of mood

The breakthrough achieved by the two researchers was immense: it showed that a large part of hormonal balance is dependent on mood, the hypothalamus being, in effect, governed by the emotions.

The discovery of the two neurohormones, TRF and LRF, to which has since been added GRF, which inhibits growth hormone, and the endorphins (see p. 000) has had another major consequence: it has fused neurology and endocrinology together in the important new discipline of neuroendocrinology. Finally, it has opened up new perspectives on psychosomatic effects which had remained uncertain until then. It has now been established, on an organic basis, that the hypothalamus is sensitive to the emotions and can influence growth, sexual function and the thyroid.

Champagne

Pérignon, c. 1680

Born of economy

The birth of champagne was entirely due to a marriage of chance and ingenuity.

The grapes of champagne, like other varieties, produce a must (unfermented grape juice) of incomplete fermentation, thus giving off carbon dioxide. At the end of the 17th century Dom Pierre Pérignon, cellar-keeper of the Abbey of Hautvillers near Epernay, noted as others had done before him, that the fizzy gas sometimes caused corks to pop in the cellars; this occurred mostly at the beginning of the fermentation, in spring.

It is difficult to be certain about this but it seems likely that the monk would first have tried to prevent this happening by keeping the cork firmly in the bottle using a strong muzzle of twine protected by covering of wax. The wine, which was always red at that time, continued its fermentation inside the bottle, no longer causing corks to pop but instead causing bottles to explode.

Since this wine delighted the palate with its lightness it quickly found favour at the court of Louis XV. The first commercial producers of champagne established themselves at Reims and Epernay. Dom Pérignon had created a new type of wine.

From red to white

It was not until the 19th century that white wine supplanted red in the manufacture of champagne (rosé champagne, obtained by adding red wine, still bears a trace of colour) and champagne became the clear wine it is today. The problem of exploding bottles was resolved well before then, though, by using thicker glass.

No other alcohol undergoes such a complex vinification as champagne. First of all the grapes are pressed; the must is placed into wine presses to settle, and remains there for between 12 and 24 hours according to its density. Then, once it is clarified this must is put into scrupulously clean barrels where it undergoes a first fermentation. Some three months later, in December, the first racking or decanting takes place and the blending which will characterize the cuvée (house style) is perfected. This is an operation demanding great skill. After racking the wine is clarified by fining then multiple filtering, racked again and laid down to rest until the summer. That is when it is put into bottles, where the fermentation continues under the cork. Six to seven months later the bottles are placed upside down, and turned periodically in order to guide the deposit which has formed towards the neck of the bottle. The bottle is then uncorked, the deposit removed and a small quantity of candied sugar solution is added, together with some wine intended to complete the characteristics of the blend; the bottle is then finally recorked and the muzzle and seal put in place.

The muzzle

The muzzle of iron wire did not replace the twine one until the 20th century, when exports of champagne were growing. It was in the 20th century as well that the coloured wax was replaced by tin foil.

Old champagne

Only a few vineyards in Champagne still produce red wine, which is characteristically richer in carbon dioxide but not sparkling; notable examples are Bouzy, Ambonnay, Ay and Damery.

Chromosomes (Function of)

Morgan, 1909

From heredity to genetics

Until the beginning of the 20th century the link between chromosomes and heredity remained hazy. Quite suddenly it became clear.

At the beginning of the 20th century biologists suspected that heredity depended on some material foundation, probably chromosomes (which had been discovered in 1888), but ideas on the nature of heredity were fragmentary.

Mendel's laws

The decisive works of Mendel (see p. 90), dating from 1865, were then totally unknown, and they continued to be so until 1907. However, a few scholars did know about them earlier. In fact, the biology of heredity (the word genetics would not be used until 1906 by Bateson) began to take shape in 1900, thanks to three botanists: the Dutchman de Vries, the German Correns, and the Austrian von Tschermak. Working independently they discovered (or rather rediscovered) Mendel's laws of plant hybridization; the Englishman Bateson and the Frenchman Cuénot applied these laws to animal biology. Heredity was shown to be governed by the mechanisms defined in Mendel's work.

In 1891 the American Henking noted, while studying the sperm of Diptera (a type of insects), the presence of a particular chromosome which was only found in half of the germ cells of the spermatozoa or spermatids; he called it X. It was also found in several other insects, and between 1901 and 1908 MacClung, Wilson, Stevens and Morrill realized that it determined the offspring's sex, according to whether it was present or absent.

In 1909 the American, T.H. Morgan, tried to establish the role of chromosomes by producing artificial mutations in the fruitfly. He had no success but nevertheless he did observe spontaneous mutations and used them in cross-breeding. He discovered heredity linked to sex or, more precisely, to the sex chromosome X. He had established a fundamental notion in genetics! Heredity is passed on through chromosomes and partly through sex chromosomes.

In 1923 several researchers were to complete Morgan's discovery: a female mammal has two X chromosomes and is said to be homogametic; a male has one X and one Y chromosome and is called heterogametic. Between 1930 and 1950 innumerable experiments in cross-breeding among a great many animal and plant species refined Mendel's laws and Morgan's discovery. Gene mapping, or at least the beginnings of it, were established. Most important, though, was the verification of the theory that sex is determined in all the cells of an individual; that is to say, chromosomes are not only carriers of sexual characteristics but also of innumerable other characteristics belonging in part to the male parent and in part to the female.

The precursors

The question of the function of chromosomes was first addressed in the 18th century by Maupertuis, then Buffon and, later, Diderot, who all supposed that 'seminal molecules' transmitted heredity. In 1868 Darwin came up with the idea that each cell can produce a seed capable of producing an identical cell.

Circular polarization

Pasteur, 1848

For life, light turns to the right

*Not all crystals polarize light in the
same direction.*

In 1848, at the age of 26, Louis Pasteur studied crystals of tartaric acid and tartrates, plant substances very widely used in food and in industry. Isolated in its pure form in 1769 by Scheele, tartaric acid had then been very closely studied by the Russian Mitscherlich and the Frenchman La Provostaye; Pasteur discovered that its crystals, like the tartrates, included small facets which only existed on half the intersections or similar angles. That is to say that the crystals are asymmetrical. Pasteur inferred from this that as the basis of this asymmetry or hemiedry there had to be a corresponding molecular asymmetry. One peculiar detail was that the crystals of paratartrate were also asymmetrical but their facets inclined some to the right, some to the left.

If Pasteur's conclusion were correct, opposite crystals of paratartrate would have to polarize light in opposite directions. To understand this, one has to know that quartz or sugar water, for example, transparent substances, have the property of making the vector or direction of light rotate because of an asymmetry, molecular in the case of sugar water and structural in that of quartz. Light can be thought of as being made up of groups of waves, all vibrating in a given direction. In natural light the direction of the different groups are randomly aligned and the light is said to be unpolarized. After passing through quartz or sugar water they are all rotating together and we have what is called circular polarization, a mysterious phenomenon in Pasteur's time.

'Mirror' structures

Under the supervision of his master Biot, Pasteur separated the right crystals of paratartaric acid from the left crystals and established that they effectively polarized light in opposite ways: the right crystals polarized light to the right (ie the direction rotated clockwise) like those of natural tartaric acid (extracted from the grape, for example), and the others polarised it to the left (anticlockwise). The left crystals were of racemic acid. Pasteur had just found the solution to a problem which the celebrated Mitscherlich had not been able to resolve: that of the composition of paratartaric acid. This was produced by the combination of two bodies with molecular structures 'mirroring' each other: natural tartaric acid, called dextrorotatory, and inorganic racemic acid, called laevorotatory. Biot had studied the same problem for 30 years without ever finding the explanation.

Following this Pasteur would demonstrate that organic acids are dextrorotatory, and inorganic ones laevorotatory, which gave considerable impetus to the study of circular polarization.

However, the great scholar had also discovered isomerism: that is, the existence of two bodies with the same chemical composition and the same molecular weight but different structures and properties (see p. 112). Nevertheless, Pasteur did not go further into isomerism, the credit for the discovery going to Boutlerov.

Circulation of the blood

Harvey, c. 1628; Power, 1649; Malpighi, 1661

'A river of blood'

All the elements needed for the understanding of the circular nature of the blood flow had been long known. When they were put together no-one believed it.

Physiology was, in the 17th century, like the rest of medicine, a body of immutable beliefs and dogmas of a supposedly philosophical nature, dominated by the teaching of the Ancients. It was believed that, as claimed by Aristotle, the blood vessels carried both air and blood and that, as Galen claimed, only the arteries contained nothing but blood, air being distributed to the veins after having entered the heart by way of the lungs. According to these two 'fathers' of science, blood was formed in the liver, passed into the right auricle (auricles and ventricles are heart chambers), and crossed the intermediary lining or septum to arrive in the left side of the heart where it was mixed with air. . . .Any kind of autopsy would have shown that there is no opening in the septum to allow any such passage.

Some progress was made with the theory of Realdus Columbus, assistant to the celebrated Vesale in the 16th century, and his pupil Andreas Caesalpinus, who put forward the idea of a distribution system for blood by the venous network; but this theory was wrong in that it stated the direction of venous flow to be the reverse of what it actually is. Another doctor, Fabricius, did discover the venous valves but did not grasp their role.

Blood circulation

This was the situation when the English doctor William Harvey, attached to St Bartholomew's Hospital, considered the question. Armed with observations and logic, he formed a theory which he published in 1628 under the title, 'On the movement of the heart and the blood in animals'. By referring to Galen he first of all rid himself of the idea that there was air in the vessels; then he demonstrated that there was no hole in the septum. He declared that the blood is expelled from the ventricles during cardiac contraction or systole, and enters the auricles during relaxation or diastole; that pulmonary circulation occurs in the pattern, right ventricle–lungs–left auricle–left ventricle; that the arterial pulse is due to the filling of the arteries; and that the venous valves serve to prevent the blood from flowing back. In short Harvey discovered the pattern of blood circulation. He missed only one element of it, but certainly not through any lack of intuitive knowledge: this was the communication of the arterial and venous systems by the mediation of capillaries (tiny blood vessels).

Harvey did not have the microscope at

> **Harvey and epigenesis**
> Studying the embryos of deer, herds of which were kept for him by King Charles I in royal parks, Harvey came to the conclusion that the origin of life was to be found in the egg and that the menstrual flow did not play a part in the formation of the embryo, contrary to what Aristotle maintained. He did agree with Aristotle on one point, though: that the foetus is formed by the uterus gradually, according to the principle of epigenesis, and not by preformation, a theory which had already been postulated at the time and still persisted, according to which the offspring develops by enlargement of a fully formed egg cell.

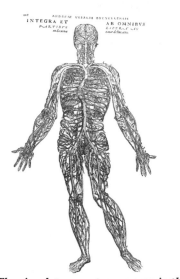

The circulatory system as seen in the 16th century *In the 16th century at the time of the anatomist Vesale, to whom we owe this etching, the placement of the blood vessels was known approximately — but only approximately as a close examination of this drawing reveals. The role of the heart was totally unknown; it does not even figure in this drawing. It was William Harvey who first discovered both the principle and the direction of blood circulation.*

his disposal, however, and it would be Henry Power in 1649 and Marcello Malpighi in 1661 who would almost complete Harvey's discovery. The former would discover under microscope the vessels which were 'fine and hair-like, linking the arteries to the veins'. The latter would encounter the same capillaries in the lungs and the mesentery (a membrane in the abdomen) of the frog; this is how it was finally understood that the blood may pass from the smallest arteries to the smallest veins (and inversely in pulmonary circulation).

Certainly Harvey's discovery was embry-

onic, for he was not able to describe the influence of the central nervous system on the circulation, nor the mechanism of cardiac contraction (Keith and Flack), nor the oxygenation of the blood by the series of gas exchanges in the lungs, nor the influence of respiration on circulation. ... but he did succeed in making a science of physiology. His pamphlet, very unorthodox in the view of some academics, excited international interest, the more so because Harvey, son-in-law of King James I's personal physician, was a doctor at the time of its publication.

Hostile reactions

Among reactions to Harvey's discovery, however, there was much hostility, hostility which was ultimately very harmful to him. The French anatomist Jean Riolan published a refutation of Harvey to which he, not being a keen polemicist, did not reply until 1649, some 21 years later. It was above all the idea of circulation, a circular flow of blood, which vexed Harvey's contemporaries. Even Descartes, while he accepted, more or less, the principle of a rotatory flow, rebelled at the idea of autonomous contraction of the heart, essential to an understanding of circulation. As far as he was concerned, expansion of the ventricles was caused by pressure on them from small intakes of blood which would be quickly vapourized by the heart's warmth. So attached was Descartes to this strange notion that he went as far as to write that should this explanation be false, his philosphy would then be worthless.

With the practice of autopsy gaining ground, and anatomy and biochemistry making progress, physiology and understanding of the circulation of the blood benefitted notably from the works of Lavoisier and Claude Bernard. Few corrections would be made to the pattern discovered by Harvey.

Coagulation (Factors of)

Henson, 1771; Hammarsten, 1876

Preserving by obstruction

The causes and mechanisms by which blood ceases to be fluid have been the object of a string of discoveries.

Nearly a century and a half after Harvey published his discovery of the circulation of the blood, an English doctor, William Henson, whose name is barely known today, was asking the logical question why, in certain circumstances including death, the blood stops circulating. He studied dead tissue under a microscope and discovered in the vessels dried blood organized in a network of fibres. This threadlike web that he was the first to see is fibrin. Nearly a century would pass without any real interest being taken in Henson's discovery.

In 1876 the German Hammarsten would establish that fibrin is formed out of the soluble blood protein, fibrinogen, under the influence of an enzyme, thrombin. In 1904 Morawitz would in turn establish that the process is more complex: thrombin owes its existence to another enzyme, then called thromboplastin and now usually called thrombokinase, which acts on a third enzyme, prothrombin, secreted by the liver in the presence of calcium salts.

Later, knowledge of the phenomenon of coagulation would be improved by distinguishing four principle factors: fibrinogen or factor I, prothrombin or factor II, tissular thromboplastin or factor III, and calcium or factor IV, each of which was the object of a separate discovery.

A therapeutic use

From the 1930s, a whole chain of secondary discoveries was to clarify our knowledge of coagulation: antihaemophilic globulin A or factor VIII (Patek and Stetson, 1937; Quick, 1947); proaccelerin or factor V (Owren, 1947), which was in fact the precursor of accelerin or factor VI; proconvertin or factor VII; plus several different factors named A to E which govern coagulation time. During the 1950s research revealed evidence of new mechanisms affecting the circulation and coagulation of the blood. Coagulation occurs only in the following three circumstances: lesion of (damage to) the tissue or vessels such as might result from cardiovascular accidents, cardiac arrest or the presence of substances, like certain toxins, which are conducive to it.

The understanding of coagulation is vital to haematology, for its causes are directly involved in blood disorders such as haemophilia. It is also of interest to a great many other medical disciplines, such as genetics, since hereditary absence of certain factors affects blood coagulation and the medicine of infectious diseases.

Coelacanth

Anon, 1938

Still around after 350 million years

To find a living fossil is rare. The discovery of this prehistoric fish was a precious find for palaeontology.

In 1938 some fishermen brought up in their nets a fish, about two metres long, caught in the Mozambique channel. Blue-grey, equipped with strong fins on stalks, it resembled no known fish. Luckily they kept it intact and when it arrived in the laboratory of the South African ichthyologist (fish expert) J.-L.-B. Smith, he was able to identify it: it was a coelacanth, a fish thought to have been extinct for 80 million years. Smith named it *Latimeria chalumnae*, the first name in honour of his secretary, Miss Courtenay Latimer, who had drawn his attention to the creature, and the second because the fish had been caught near to the mouth of the river Chalumna. The discovery aroused a great deal of interest, because it was exceptional, and also for scientific reasons. A closely related creature, *Malania anjouanae*, found in 1952 near Anjonan in the Comoro Islands in the Indian Ocean, caused a similar stir.

Belonging to the family of the Coelocanthids of the class of osteichthyes and the subclass of Crossopterygii, bony fish appearing in the Devonian period 350 million years ago in the gentle waters of Spitzberg, the coelacanth has enabled a certain number of suppositions to be verified. In the first place it confirmed that evolution does not always mean progression; it may also mean regression. The skeleton of the coelacanth is a step back

60 coelacanths
About 60 coelacanths have been found and each discovery is a special event.

Coelacanth A living fossil, a title shared with the shark and cockroach, for example, the coelacanth was of particular interest, supplying as it did much information about the Crossopterygii, fish which appeared in the Devonian period and which are at the origin of all the four-footed vertebrates. Since 1938 around 60 of them have been caught.

from the secondary coelacanthids, for it is very gristly. Furthermore, the brain is clearly smaller than the capacity of the cranial cavity, which also suggests regression. One other point should be made: evolution is not inevitable. In some 80 million years the coelacanth has not altered.

A living paradox

As well as signs of regression the coelacanth retains characteristic evolutionary traits, such as the peduncules (stalks) of its fins, which make it, like all the Crossopterygii, an ancestor of the vertebrates; its fins denote a capacity to move both on land or in water, being both fins in the traditional sense of the term, and embryonic limbs of locomotion.

One peculiarity of the coelacanth which is of particular importance is the degeneration of its respiratory system. While the rhipidistians, another family of Crossopterygii related to the coelocanthids, possessed a dual respiratory system, using both gills and lungs which allowed them to breathe equally well in air or in water, the coelacanthids lost this faculty. It was, however, the rhipidistians which disappeared and the coelacanth which survived.

Coffee

Kaldi (?), c. 850

Insomniac goats

---

*The discovery of the coffee plant
remains obscure despite the precise
nature of the anecdote.*

Around the year 850 a young Ethiopian goatherd, Kaldi, was astonished to see that his goats were not sleeping at night. Watching them closely, he discovered that they were grazing on the red fruits of an unknown tree. Having tasted the fruits, the goatherd in his turn found that they had a stimulating, even an exciting, effect on him. The legend is improbable for the berries of the coffee tree are bitter. However, the drink called coffee, a name perhaps derived from the Arab word *qahwah* or alternatively from the name of the province Kaffa, in Ethiopia, where coffee is cultivated, was indeed discovered around the time mentioned in the legend. One other point is certain: all coffee trees, which belong to the genus *Coffea* in the *Rubiaceae* family, originated in Ethiopia. Coffee is an evergreen shrub whose cherry-like fruits contain the coffee beans. Complete plants were not exported until the 15th century, destined for Arabia. Coffee houses flourished in Mecca, then in Egypt and Turkey. Coffee was later introduced to the E Indies, W Indies, S America and Africa. Coffee was introduced to the UK in the 17th century, and in 1652 the first coffee-house opened in London. Coffee-houses flourished at the same time on the continent of Europe and in the USA. It is now drunk in most countries of the world, its popularity enhanced by the introduction of instant coffee by the Swiss firm Nestlé in 1937.

Caffeine

The effects of coffee seem to be due to its caffeine content which differs according to variety (the strongest being robusta), industrial production and method of domestic preparation. Caffeine is an alkaloid, a weak stimulant of the central nervous system, found in tea and cocoa as well as coffee.

Cortisone

Kendall, 1935

Powerful therapeutics

This corticosteroid hormone,
infinitely useful in numerous
treatments, was to be long neglected.

Since 1855 medicine had known the classi-fication of Addison's disease, named after the doctor who first described it. This serious illness is caused by the bilateral destruction of the adrenal glands (usually followed by tubercular lesions) and notably produces a loss of water and salt, low blood pressure, tiredness, hormone deficiency and its various consequences, and wasting. One characteristic symptom is the tanned look of the skin, which gives rise to its sometimes being called 'suntan disease'.

The forerunners

In 1856 the Frenchman Brown-Séquard established that the adrenal gland is indis-pensable to life. In 1920 the German Biedl established that this was also the case with the covering or cortex of this gland. The theory was tested on animals in 1927 (Rogoff and Stewart): after removing the adrenals, extracts from the glands were administered to the animal. These cry-stallized extracts were very active, but while it was known that they were steroid in nature there was still a lot to learn about them. In addition, the isolated substances were complex and the specific role of each constituant was unknown.

In 1935 the American Edward Calvin Kendall and his colleagues, working on the subject in the famous Mayo clinic in Roch-ester in the USA, using adrenals from cattle, isolated an unknown substance which Kendall called 'compound E' with a specific formula. Two years later he isolated

another specific substance, 'compound F'. The intrinsic properties of these compounds were still unknown, just like those of other corticosteroid hormones discovered at around the same time: dehydro-corticosterone, isolated in 1935–6 by the same Kendall with Mason and Edward; corticosterone, also identified by Kendall in collaboration with Mason and Reichstein; adrenosterone, isolated in 1937 by Reichstein; and desoxycorticosterone, iso-lated and partially synthesized in 1937 by Reichstein, von Euw and Steiger.

The nominal objective was still simply the treatment of Addison's disease but many other applications were indicated and a lot of research took place. Without knowing exactly what use it could be but knowing intuitively, guided by many exper-iments, that 'it could be useful', particular attempts were made to synthesize these sub-stances whilst discoveries of corticosteroids proliferated. In 1945 Reichstein succeeded in the synthesis of corticosterone which had been discovered meantime. In 1946, Sarrett and, in 1947, Kendall, in turn achieved synthesis of 'compound E', from then on

---

Cortisone and stress
Like other corticosteroids, cortisone, a glucocorticoid which proceeds from the metabolism of cortisol and which is metabolized in the liver, can undergo important variations on account of physiological and psychological problems brought together under the heading of stress.

christened cortisone and with a formula summarized in the scientific name, 11-hydroxy-17-corticosterone; compound F had meanwhile been renamed hydro-cortisone.

No-one suspected the real value of these discoveries, for all the researchers were seeking another corticosteroid. This was aldosterone, which was not to be found until 1952. Synthesis of this hormone, 50 times more active than desoxycorticosterone, was to be achieved three years later.

On 13 April 1949, Kendall, Hench, Polley and Slocumb reported to the International Congress of Rheumatic Diseases observations made on 16 sufferers of chronic developing polyarthritis who had under-gone injections of 100 mg of cortisone a day for several days: the signs of the disease had disappeared! Excitement over this was intense, all the more so since the same results could be achieved by injections of another hormone, this one secreted by the pituitary gland: adrenocorticotrophic hor-monic or ACTH. It was thought that not only Addison's disease but all rheumatic diseases had been conquered.

The use of cortisone

The range of serious illnesses for which corticosteroid therapy is used is still growing:

Addison's disease
chronic polyarthritis
spondyloses
psoriatic arthritis
acute rheumatism of the joints
systemic lupus erythematosus
scleroderma
dermatomyositis
polyarthritis nodosa
most collagen diseases
haemolytic diseases
myeloma
thrombocytopenia purpura
certain cirrhoses of the liver
lipoid nephrosis
acute tubercular infections
numerous allergies

Hench's intuition

Observation had been joined by luck in this therapeutic rediscovery of cortisone. Hench had noticed that in the course of pregnancy or after jaundice of the liver, polyarthritis seemed to be cured. He reckoned that this was a definite indication of the action of a hormone operating in both sexes and linked to the bile acids. Now it was precisely from desoxycholic acid that Sarrett had achieved synthesis of cortisone. Kendall, therefore, advised Hench to administer cortisone to his patients, with well-known results. This is how Kendall can be said to have discovered cortisone twice.

The dangers of cortisone

Experience was to reveal important side-effects of the hormone: retention of lipids, water retention, loss of potassium, dim-inution of white blood cells (lymphopenia), and lowering of resistance to infection. The action of cortisone on the cortico-adrenals themselves is not negligible since it can lead to either an excess of or a reduction in cortical secretion. Equal account must be taken in corticosteroid therapy of the risks of perforating digestive lesions, thrombo-embolitic accidents, shocks and psycho-logical problems. Since the 1960s it has been proved that this treatment must be reserved for serious illnesses.

Corticosteroids

Cortisone belongs to one of three groups of corticosteroid hormones which are defined by their chemical structures; this group is that of the 11 oxycorticosteroids. The second group is that of the 17 cetosteroids, hormones with a sexual action also secreted in men by the testicle (but containing even in men, a weak element of oestrogens). The third group consists of a simple hormone, aldosterone.

The corticosteroid hormones regulate the metabolism of minerals, fat, nitrogen and glucose.

Cosmic background radiation

Penzias and Wilson, 1965

Remnants of the Big Bang

From the explosion which created the Universe, there remains a background of electromagnetic radiation, the same in all directions

In the Big Bang theory of the universe it is envisaged that all matter and energy now observed in the universe originated from a very compact, infinitely dense, extremely hot fireball. Present-day estimates are that this primordial explosion took place 20 thousand million years ago. In 1948 proponents of the Big Bang theory such as Ralph Alpher, Robert Herman and George Gamow made an important prediction. When the early universe had cooled sufficiently for the protons and electrons to combine to form hydrogen atoms it became a transparent gas instead of an opaque hot plasma. An electromagnetic radiation field characteristic of the temperature at which combustion takes place expands for 20 thousand million years and cools to a very low temperature. This radiation field would then be the same in all directions and be about 10 kelvins (10 degrees above absolute zero).

Some 15 years later, in 1965, Arno Penzias and Robert Wilson made a discovery of great importance. At the time these scientists from Bell Telephones were using a radio telescope to try to detect signals from two of the early communications satellites, Echo I and Echo II. Unlike modern satellites which receive, amplify and re-transit radio signals, the early versions merely reflected any radio signal striking them.

Penzias and Wilson were observing in the microwave region (radio waves with a few centimetres' wavelength), and because signals were weak they had to reduce all sources of radio noise interfering with the measurement. However hard they tried they could not get rid of a faint hiss which came equally from all directions in the sky and did not vary with the kind of day or from day to day. This noise was characteristic of a body at a very low temperature of 3 K.

At about the same time Robert Dickie, P. J. B. Peebles, David Roll and David Wilkinson at Princeton University generated a prediction using their own hot big bang model for the origin of the universe. The Princeton group calculated a temperature of a few degrees for the background radiation and also that it should be observable with radio telescopes. Ideas from the theoreticians and observers combined to suggest that the observations of Penzias and Wilson were the 'Birth pangs' of the universe.

This discovery of the cosmic microwave background radiation is the most important one in cosmology since Hubble's discovery of the expansion of the universe. Since the discovery many measurements at microwave millimetre and infrared wavelengths have been made from ground-based telescopes and spacecraft. A more precise temperature of 2.7 K has been determined. The strength of cosmic microwave background has been found to be the same in all directions to one part in ten thousand.

Cro-Magnon and Neanderthal men

Anon, 1856, 1868

The first Europeans

*A little more than a century ago
man's evolution by stages became an
accepted fact.*

Until the middle of the 19th century the idea of a physiological and psychological evolution of man since his origins, and especially since animal origins, was mere speculation. Palaeontology (the study of fossils.) was in its infancy and there existed very little hard evidence of this sort of evolution; what was available — the stone tools exhumed near Abbeville by the Frenchman Boucher de Perthes for instance — was rejected by many authorities, including of course the religious ones, as presumption. Charles Darwin had not yet constructed his great theory of evolution. In most universities creationist theories predominated, and as they are based on the Bible, they hardly left room for any scientific proof.

The first *Homo sapiens*

It was against this background that in the summer of 1856 workmen in the Neander valley, a steep gorge near Düsseldorf in Germany, accidentally uncovered a cave some 18 metres above the riverbed of the Neander, a tributary of the Rhine. Digging out blocks of rock to extract limestone, they discovered old bones which they didn't bother much about: most of them were broken and lost, and it was by an extraordinary fluke that a skull and some bits of skeleton were retrieved and delivered to the scientific community. The low arch of the brow, the projecting lower jaw, the extreme bulging of the forehead — all characteristics reconstructed from the skull — and the heavy posture reconstructed from the skel-

etal remains all combined to provoke a horrified reaction, even among the scholarly world, who refused to see in this a possible ancestor of humanity. One professor, F. Mayer of Bonn, even claimed that these were the remains of a Mongolian cossack who had taken refuge in the cave while on the way to Prussia during the pursuit of the Napoleonic army in 1814! Mayer even suggested that the unfortunate man suffered from congenital rickets which had deformed his face. ... It must be said that the rest of the scientific community were hardly more enlightened and many scholars spoke of a 'representative of a savage race' or a 'congenital idiot', as recalled by Professor Richard Leakey, the celebrated anthropologist, in *The Origins of Man*. Nearly half a century was to pass before Neanderthal Man was given his rightful

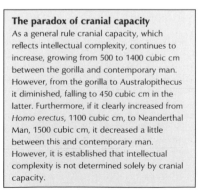

The paradox of cranial capacity

As a general rule cranial capacity, which reflects intellectual complexity, continues to increase, growing from 500 to 1400 cubic cm between the gorilla and contemporary man. However, from the gorilla to Australopithecus it diminished, falling to 450 cubic cm in the latter. Furthermore, if it clearly increased from *Homo erectus*, 1100 cubic cm, to Neanderthal Man, 1500 cubic cm, it decreased a little between this and contemporary man. However, it is established that intellectual complexity is not determined solely by cranial capacity.

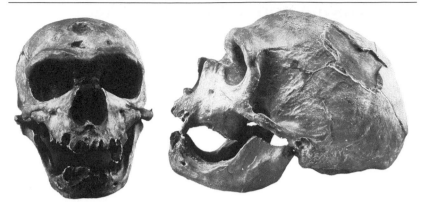

Skulls of Neanderthal men (La Chapelle aux Saints)

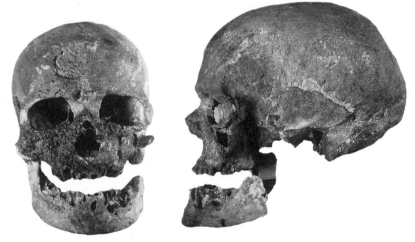

Skulls of Cro-Magnon men (Les Eyzies)

place. He is one of the first examples of *Homo sapiens*; he existed in the period between 20 000 and 100 000 years BC. Given such a climate it can be understood why the first Neanderthal-type man, actually discovered in Gibraltar in 1848, was of no interest to anyone.

In 1868 another team of workmen, clearing the way for a railway across the neighbouring cliffs of Les Eyzies in the Vézère valley in the Dordogne, came across another cave called Cro-Magnon. There they discovered five human skeletons with elongated skulls and small jaws, features of present-day man. This time the discovery was given a better reception and interpretation. It was suggested that Cro-Magnon Man was a contemporary of Neanderthal Man. It is now known that they co-existed for around 20 000 years, Neanderthal Man becoming isolated in his limited territories and eventually disappearing. Many other subsequent discoveries would consolidate and illuminate the different stages and circumstances of human evolution. Firstly other fossil men (Ehringsdorf and Steinheim in Germany; Krapina in Yugoslavia; Saccopastore in Italy, Gibraltar, and then in Palestine) were to reveal the existence of a pre-Neanderthal type, whose different representatives are grouped together as Palaeanthropians (the earliest variety of man). It has since been shown that Neanderthal Man and Cro-Magnon Man have developed from the same family, Cro-Magnon evolving towards modern *Homo sapiens sapiens* and Neanderthal disappearing without descendants.

On the way to evolution

The reconstructed physiognomy of Neanderthal Man has long made him appear to have been a brute with no feelings. However, the uncovering of the grotto of Shanidar in Yugoslavia, where an aged individual was buried, indicates at least that Neanderthal Man practised elaborate and touching funeral ceremonies; the skeleton found in this grotto had been laid on a bed of leaves and his grave decorated with armfuls of flowers. Other indications also suggest a relatively complex and evolved social organization.

Homo erectus

Other discoveries — such as that of Java Man, discovered by the Frenchman Dubois in 1891 and since renamed *Homo erectus*; that of *Sinanthropus* discovered in China between 1922 and 1940; and that of Atlanthropus, discovered in Algeria in 1954 and which lived 500 000 years ago — revealed that evolution does not occur in a uniform manner everywhere in the world and that the stage of *Homo erectus* preceded that of the Neanderthals and doubtless co-existed with the pre-Neanderthals. Right at the beginning is found a group which is not yet of the *Homo* genus but which is also no longer ape-like: this is Australopithecus (see p. 25), the first specimen of which was discovered in Africa in 1925. It was the size of a chimpanzee but stood upright; its cranial capacity was a third of Neanderthal Man's, but its dental formation was somewhere between that of an ape and that of later stages. Still further back will be found the great apes like Ramapithecus or Aegyptopithecus. Since the beginning of this century the case for evolution has been accepted and it is now only the mechanics of it, general and local, which are debated.

DNA (Structure of)

Crick and Watson, 1953, 1961

The ladder of life

Knowledge of the architecture of deoxyribose nucleic acid has been crucial to the development of molecular biology.

In 1868 the German, Miescher, took some pus cells, isolated the nucleus, and then extracted from that nucleus a substance which he called nucleine, later to be called nucleic acid. In 1874 Miescher did in fact isolate nucleic acid. Between 1882 and 1897 Kossel set about analysing this acid and discovered that it consisted of phosphoric acid, a nitrogenous base and sugars, structured in elementary units called nucleosides. A nucleoside is always formed from a nitrate base, purine or pyrimidine, and a sugar, either ribose or deoxyribose. The work of these pioneers was known only to specialists; they succeeded, however, in opening a fundamental chapter in genetics. Since 1920 it has been known that when the sugar of a nucleoside (changed to nucleotide by attaching a phosphate to it) is a ribose then one is dealing with ribonucleic acid or RNA and when the sugar is deoxyribose one is dealing with deoxyribose nucleic acid or DNA.

The beginning of cytology

Furthermore, after 1924 it was known that DNA only existed inside the nucleus of the cell and that it was the constituent of the genetic chromatin (so called because of its extreme sensitivity to basic colorants) or euchromatins which are found in the genes of chromosomes. It was also known at the time that DNA was the means of genetic transmission and that it maintained life in the cells and indeed in the entire organism. No life, therefore, is possible without DNA. As for RNA, its presence in both the nucleus and the cytoplasm surrounding it had been known since 1940. The specific roles of these acids, however, were not yet known. Finally, it was known that specific enzymes (see p. 187) activated the two acids: ribonuclease for RNA and deoxyribonuclease for DNA. Cytology (the study of cells) had indeed got underway.

Scientific emotion

The first biologists had, however, made some mistakes. They had held for a long time that DNA and RNA were in some way enclosed within the cell. Then in 1944 the Americans, Avery, McLeod and MacCarthy, discovered to their amazement that the DNA of dead bacteria was able to modify the genetic material of living organisms (see

---

An important date

The first major date after Miescher's discovery was 1944, the year when the Americans Oswald T. Avery, Colin McLeod and Maclyn MacCarthy demonstrated that the genetic material of all living beings is made up of DNA.

---

p. 79). This transmission of genetic information from one organism to another outside the process of fertilization caused a stir in the biological community. Curiously enough there were two celebrated German physicists, Erwin Schrödinger and Max Delbrück, who, in the 1940s, were to give a decisive boost to our knowledge of DNA. Schrödinger suggested in a book with the simple title *What is Life?* that there were real biological mechanisms which were still incomprehensible, but he assigned a major role in these yet-to-be-discovered mechanisms to the gene.

The genetics of viruses

Delbrück, who had been working for several years on the genetics of viruses which can infect bacteria (called bacteriophages), gave a series of ground-breaking lectures in New York in the summer of 1945. He in effect set out the major lines of genetic transmission; in infecting the bacteria the virus inserts its own DNA into that of its host. This modified DNA will not only produce its own proteins but will also be used in the replication of the viruses whose code it carries. This translation of DNA is brought about by the relay of different nucleotides which constitute the DNA. These lectures were widely disseminated and reached an international public among geneticists.

Delbrück was by no means the only one to have read Schrödinger's book. At least two others had done so: the New Zealander Maurice Wilkins, also a physicist (he collaborated in the making of the first A-bomb) but who converted to biology; and Max Perutz, a scientist of Austrian origin, living in Britain and a specialist in the diffraction of X-rays. In the years after the end of the war Wilkins and Perutz found themselves in the molecular biology department of the Cavendish Laboratory in Britain under the direction of a pioneer: Lawrence Bragg, son of William Bragg. From 1947 Lawrence Bragg, with the German Max von Laue, succeeded in perfecting diffractometry by X-rays, the principle of which is as follows: the X-rays are diffracted by the atoms of a crystal and the resulting 'diffraction pattern' gives information about the atomic structure of the

Diagram of DNA
This diagram illustrates both the structure of DNA and its double spiral (upper section) and also the way in which it divides in cellular 'mitosis', each of the spirals generating its opposite partner by production of the original corresponding bases.

crystal. The essential point to note is that Wilkins and Perutz extended this technique to the study of proteins.

Nearly all the strands were being drawn together in readiness for an attempt at the problem which intrigued all the experts: what is DNA made up of and how is it laid out? One final weapon was then added to the arsenal: chromatography on paper, perfected by another great pioneer, Edwin Chargaff, a naturalized American of Austrian origin. Thanks to this technique it became possible to measure the proportion of bases in nucleic acids.

The structure of DNA

Using the data already gathered by Chargaff and the method of Wilkins and Perutz, Francis Crick and James Watson, two young researchers (the former English, the latter American), concentrated their efforts on DNA. They discovered 'a rope ladder wound in a spiral'. Geometrically it is a double spiral or helix held by 'rungs' which are formed by nitrogenous bases, purine and pyrimidine. The backbone of each of the spirals is formed alternately by molecules or nucleotides of a phosphate (orthophosphate) and a sugar (deoxyribose) — as the Englishman Alexander Todd had indeed demonstrated in 1944 — plus one base always chosen from among these four: adenine, guanine, thymine or cytosine. Each nucleotide therefore consists of a segment of the backbone and half a rung. The particular nature of this structure is that adenine only unites with thymine, and guanine with cytosine. Crick and Watson published their discovery in 1953. It would subsequently be revealed that the matching of bases was largely the result of Watson's researches, while the synthesis which culminated in the three-dimensional model was due mainly to Crick.

The role of intuition

The role of intuition in scientific discovery is particularly evident here. Chargaff had noted that the four nitrogenous bases of DNA appeared in a regular fashion in chromotography. The adenine combination (A) appeared as often as the thymine combination (T), and the guanine combination (G) as often as the cytosine combination (C). In short there were two sets of equivalences: A and T on the one hand, G and C on the other. Watson cut out of cardboard the groups A, T, G and C whose structure was known (they were held in place by an atomic bond between hydrogen atoms), and played around with his model trying to work out a structure. 'It was then', wrote Watson, 'that I noticed that an A–T pair held by two hydrogen bonds had the same form as a G–C pair.' He had just discovered that the rungs could follow each other naturally. It only remained to construct the three-dimensional model of DNA. In this the bases follow each other in this order: AT, GC, TA, GC, AT, AT, GC, CA, the formula varying according to whether one of the bases is situated on the left or on the right of the double spiral.

Francis Crick carried on with his research and in 1961 established that each group of three bases or 'triple', on just one of the two strands of DNA, codes for each of the 20 amino acids normally found in a protein. Watson too, at the same time, established the role of ribonucleic acid or RNA in the synthesis of proteins: it is a 'mould' of DNA which transmits the instructions of synthesis.

Replication of DNA

The reverberations following the discovery were immense although it was in effect only the culmination of years of research. At last we had the original alphabet of living matter: four bases could combine, in different ways, to form the 20 amino acids, whose combinations in turn formed the different proteins of the cell and of all living tissues. The actual method of duplication of DNA in each cell was still to be determined. Three models were proposed. One was the separation of the two helical strands of DNA, each of which would come together with a new series of nucleotides; the second was the production of nucleotides without rupture of the double helix; and the third proposed dividing the double helix into triplets which would replicate themselves separately. The three hypotheses were tested and the first was the best. At a given moment in a cell's life the double helix splits into two separate helical strands which, trapping the sugar, phosphate and base in the middle, reconstitute their components. The law of the matching of bases ensures that the reconstitution is exact. The joining between the phosphates is achieved through the agency of an enzyme, DNA-polymerase. The process is progressive; it is almost as if each edge of a sliding door which is slowly being opened reconstitutes its opposite side as it goes along.

Linus Pauling's mistake

Chemist of high renown (he received a Nobel prize) and Crick and Watson's professor, Pauling might perhaps have found the structure of DNA at the same time as his former students, if not before, if he had not followed a false scent: he thought that this collection of genetic information was a protein, when it is in fact an acid.

Three Nobel prizes

The discovery by Crick and Watson is one of those which appear to the scientific historian to be inevitable. Without wishing to detract from their achievement, it must be noted that these researchers had at their disposal most of the elements necessary for their discovery. This was thanks to the work of others, in particular Wilkins and Rosalind Franklin who had taken photos of DNA. The illustrious chemist, Linus Pauling, also pursued this line of enquiry and though he did not manage to conclude his researches, it is very probable that Pauling's example also contributed to Crick and Watson's success. Pauling was in fact their professor and he had already discovered, in the field of organic chemistry, the structure of the alpha spiral. Pauling's inspirational role was, moreover, recognized by the Nobel jury who, in the same year that it honoured Crick and Watson, awarded him the Nobel prize . . . for peace!

An unrecognized contribution

The acclaim given to Crick and Watson made them household names. Justice demands that mention should also be made of Rosalind Franklin, the crystallographer who, in collaboration with Wilkins, perfected the techniques of crystallography as applied to the study of proteins. Her work was essential to Crick and Watson's discovery. Unfortunately she died at the age of 37 in 1958, four years before the Nobel prize was awarded to the two researchers.

The spin-offs

The fact remains that knowing the structure of DNA set molecular biology off on the trail of its great achievements. In addition to its innumerable benefits to biology and virology (it helped to establish how viruses infect cells), its contribution to genetic engineering must be stressed. In 1969 the Americans, Lawrence Eron, James Shapiro and Jonathan Beckwith from the Harvard Medical School succeeded in isolating for the first time a single gene, the lactose gene, from among the 3000 or so genes from the bacterium *Escherischia coli*. The following year the Indian Har Bind Khorana made the first successful synthesis of a gene consisting of 77 pairs of bases, that of alanine, an amino acid, and did it by taking as a basis the structure of the corresponding RNA.

Correcting genetic defects

Knowing the laws of the matching of bases, biologists can now modify at will the DNA of numerous living organisms in such a way as to have them synthesize a particular protein. They can also reconstitute DNA integrally after taking an inventory of the nucleotides and their composition. Finally they can envisage in the not-too-distant future being able to correct genetic defects in a living being by extracting cells, modifying their DNA and grafting these cells on to their subject again.

Doppler–Fizeau effect

Doppler, 1842; Fizeau, 1848

From the tuba to the planets

The frequency of a sound wave changes according to where the observer is situated. Music lovers know this instinctively.

In 1842 Johan Christian Doppler, an Austrian, one of the most celebrated mathematicians and physicists of his time, noted that the frequency of a sound as it is perceived by a listener in a fixed position (legend has it that he grasped the phenomenon thanks to a type of tuba) varies as the source approaches and retreats, becoming more shrill in the first case and lower-pitched in the second. Doppler extrapolated this phenomenon to light, and expanded it theoretically in a small treatise of astronomy entitled, 'On the coloured light of double stars'. Doppler had grasped the whole import of his theory, for the discrepancy of frequency can, if the observer is in a fixed position, modify the wavelength of light or sound emitted by the moving source and, if the observer himself is in motion, modify the apparent speed of the source.

In 1848 the French physicist Armand Fizeau rediscovered this independently; it is understandable that he hadn't heard of Doppler's work, for scientific communication was less rapid then than it is now, and so the phenomenon is often referred to as the Doppler–Fizeau effect.

Other 'effects'

Use of the Doppler–Fizeau effect has long been extended to a number of areas from astronomy and astrophysics to sonars and radars, from directional maritime and aerial systems to nuclear physics. It was given an extended life in the nuclear domain by the discovery of a similar phenomenon, the Mössbauer effect, discovered in 1958.

One of the most remarkable applications of the Doppler–Fizeau effect in astronomy has been the possibility of calculating the speed of rotation and distance of celestial bodies. Thus in 1965 the Americans Dice and Petergill succeeded in measuring the rotation period of Mercury in the following way: they sent radar waves towards the planet with the intention of picking them up on their return. Using the Doppler effect of Mercury's rotation (one side of the planet is moving away from us, and the other moving towards us) on the radar waves, gives the rotation period of the planet as 59 + or − 3 days. (These radar emissions were repeated hundreds of times over the months.)

Indispensable to the verification of Einstein's theory of relativity, the Doppler–Fizeau effect has helped to establish that light from a heavenly body is shifted towards the red (longer wavelength) end of the spectrum when the star and the Earth are moving away from each other by the 'stretching out' of the waves, and towards the violet (shorter wavelength) end when the star and the Earth are approaching each other.

The Big Bang

The shifting towards towards the red (or 'redshift') of the light from the vast majority of galaxies, was the foundation for the theory of the Big Bang which claims that the universe began with a huge explosion and is constantly expanding. It is thanks to the Doppler–Fizeau effect that the dynamics of the Milky Way (our own galaxy) can be understood and that we can grasp the nature of quasars, enigmatic objects with very small optical images and with very pronounced redshifts which indicate that they are moving very rapidly away from us.

Electric battery

Anon, c. 250 BC; Galvani, 1780; Fabbroni, 1796; Volta, 1800

Bottled energy

The accumulation of electricity in an object of reduced volume is perhaps much older than had been thought.

The classic view is that the electric battery was an invention of the Italian, Volta. The history of science records that the great scholar demonstrated his apparatus to the Academy of Sciences in 1801 and then to the Institute of France, in the presence of Bonaparte, who awarded him a gold medal.

An ancient battery

This point would therefore seem to be straightforward and unmysterious, were it not for the fact that in 1957 an object was examined which is now in the archaeological museum of Baghdad in Iraq, and a diagram of which is reproduced opposite. Dating from 250 to 244 BC, that is to say to the era of the Parthian occupation of the Baghdad region, this artefact, the size of a modern pocket torch, seemed to be nothing other than an electric battery. All that was lacking was a conductive wire, going from

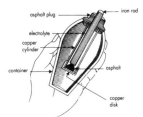

Electricity 22 centuries ago *The ancient object of which this is a diagram and which is now in the archaeological museum of Baghdad could not logically be identified as anything other than an electric battery. However, it remains rather odd that no other example has been found of this object, which could perhaps have been used in gilding by electrolysis.*

the copper cylinder to the exterior, which could perhaps have become separated over the centuries. The object seemed perfectly authentic, even though an iron-copper battery from the 3rd century BC would be quite amazing.

Identified by the German archaeologist Wilhem König in 1957, the battery was tested by the American Willard F. M. Gray of General Electric's High Voltage Laboratory, who researched the type of electrolyte chosen for the working of such a battery; he did several experiments on it until he decided on copper sulphate. Gray also thought that acetic acid or citric acid, which was readily available at that time, would also have made an excellent electrolyte. In any case, once the electrolyte was added, the battery — or rather an exact model reconstructed from it — functioned perfectly.

What use the Parthians could have made of an electric battery is unclear. Given that they had no lamps, it could only have been of limited use. The most probable use is electrolysis (the application of a thin coat of metal to a metal surface), used in silver-working, for example. It is a known fact

---

Much earlier

Copper and iron existed in the 3rd century BC, and it is quite possible that, having hung a slaughtered animal from an iron bar, the Parthians might have noted the animal's contractions. Then, taking their observations further, they might have discovered that copper and iron plunged into a container of acetic acid produced the circulation of a current. It would then follow that the electric battery was created as a result of a discovery made long before Galvani's experiment which inspired Volta.

that much antique jewellery is not solid silver or gold but simply covered by extremely fine leaves of precious metal. In many cases it has been established that the fineness of the layer of precious metal has been obtained by hammering on to the object itself, which is made of copper or bronze, for example. The object is then heated in such a way as to obtain the perfect adherence of the precious metal; but it is possible that, in certain cases, the silvering may have been achieved by electrolysis.

By the 3rd millennium BC the jewellery craftsmen of very many ancient civilizations had attained a level of technical perfection which greatly surprised historians and archaeologists. Soldering was widely practised in the 3rd millennium and in Mexico jewellers used complex chemical baths to rid gold-copper alloys of traces of copper from the surfaces.

Questions on the past

Perhaps electricity could have been discovered in the 3rd century BC, if not before, in circumstances similar to those in which the Italian Galvani discovered, in 1780, what he called 'animal electricity'. He attached a copper hook to the spinal cord of a frog which he was dissecting and hooked it on to some iron netting. The frog was seized with spasms just as it had been when the researcher touched the animal's crural (leg) nerve with his scalpel.

Quite a collection *In order to test the effects of the Volta battery, the Englishman Wollaston produced, at the beginning of the 19th century, this collection of 2000 batteries.*

Galvani did not understand that the frog's spasms were due to the difference of electric potential between the two metals. Keeping in mind the fact that the same spasms had been triggered when the animal was in close proximity to an electric machine (and therefore within an electromagnetic field), he latched on to the idea of a mysterious animal electricity.

Chemistry and electricity

Galvani's fellow countryman Volta was better equipped to understand the phenomenon, since he had invented several pieces of electric apparatus, including a much improved version of Franklin's glass pane electrometer (a device for measuring differences in potential). He therefore grasped the implication of Galvani's experiment immediately; it was equivalent to the one which could be done by placing the tongue between two discs of different metals connected by an electric wire, as the Swiss Sulzer had already noted in 1754.

Animal electricity does not exist, there is only electricity pure and simple. In 1796 a true forerunner, Fabbroni, also a compatriot of Volta, had discovered that if two strips of different metals were placed in water, touching each other, one of them was oxydized. This was a genuine chance to trace a link between a chemical and an electrical reaction. All that remained was for Volta to invent his battery: it consisted of a series of pairs of zinc-copper discs in direct contact, isolated from one another by moist cardboard.

It was Davy who, in that same year, was to explain the role of the liquid, the electrolyte: it had to be naturally able to oxydize one of the metals. Volta had not worked it out, but what about the Parthians? Did they know?

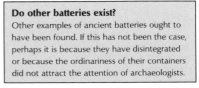

Do other batteries exist?
Other examples of ancient batteries ought to have been found. If this has not been the case, perhaps it is because they have disintegrated or because the ordinariness of their containers did not attract the attention of archaeologists.

Electroencephalogram

Caton, 1875; Berger, 1924; Neminski, 1925

The waves from within

For a long time the possibility of picking up and transcribing the electrical activity of the brain was held to be illusory.

At the end of the 19th century numerous researchers set themselves to study the role of electricity in physiological functions. To this end they applied electrodes, a little haphazardly, to the human body. Thanks to electrodes applied to the skull, in 1875 the Englishman Richard Caton detected electric activity. He registered them but no-one paid any attention.

In 1924 the German, Hans Berger, professor at Jena University, produced the first true electroencephalograms (EEGs), which were interesting in that some had been performed on subjects suffering from cerebral lacuna (small areas of cerebral ischaemic infarction) and others on normal subjects, which allowed for a comparative study. It emerged from this that there existed some constants in the EEGs. Berger's colleagues regarded him as a crank and took little or no interest in his work. The honours were taken away from him for a time by the Russian, Neminski, who also detected definite electrical frequencies in what he called 'electrocerebrograms'. However, in the works of Berger, published in 1929, data of the first order including that of alpha rhythms of 8–11 Hertz (cycles per second) is recorded. Due to the quality and detail of Berger's researches he is generally considered to be the true discoverer of the electrical potentials of the brain. That is doubtless due in a large part to his 'rehabilitation' by the Englishmen Adrian (Nobel prizewinner) and Matthews who, having verified and confirmed his work, made known its worth in 1934.

Rapid progress

After that, discoveries came thick and fast: from 1935–7 the Americans Loomis and Davis studied EEGs recorded during sleep and demonstrated that this led to a modification of the trace. Then it was discovered that the EEG of an adult is different from that of a child (1936); that anoxia (lack of oxygen), hyperventilation and hypoglycaemia (low blood sugar) (1936, 1938 and 1941 respectively) also modify the trace; and that EEGs enable us to detect abscesses and tumours of the brain (1936 and 1937) as well as the after-effects of cranial damage (1942). ... It was only in the 1940s that the EEGs became widely used in neurology.

Cerebral wave traces

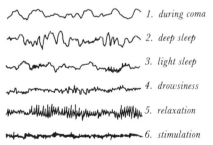

1. *during coma*
2. *deep sleep*
3. *light sleep*
4. *drowsiness*
5. *relaxation*
6. *stimulation*

Rest and sleep

A person at rest produces waves at a mean frequency of 10 Hz, called alpha waves. If the eyes are opened the trace varies according to intellectual activity and emotion. Sleep is characterized by slow waves of 2–3 Hz, often interrupted by rapid waves which coincide with dreams, called paradoxical sleep. The Lyon School and Professor Jouvet have revealed the existence of traces particular to dreams.

Electromagnetic induction

Faraday, 1821, 1831

The intuition of a genius

The possibility of converting magnetism into electricity was one of the greatest discoveries of physics.

One of the most inspired examples of intuitive thinking in the whole history of physics was to grasp that electricity and magnetism were closely linked. It was the achievement of the Englishman Michael Faraday, a genius who was without doubt physics' greatest experimenter. It culminated in one of the most fruitful of scientific discoveries: electromagnetic induction. In 1821, one year after Ørsted's discovery of the induction of a magnetic field by an electric current (see p. 123), Faraday strove to extend knowledge of the links between electricity and magnetism. He discovered that an electric current exerts a force on a magnet and, inversely, that a magnet exerts a force on a current-carrying wire. This discovery, which came very close to electromagnetic induction and merits a mention as a preliminary stage, was achieved with the help of disarmingly simple equipment: a magnet was half-plunged into a vessel of mercury, a single pole emerging from it and the other remaining fixed in the axis of the equipment. A current-conducting wire came from the base of the vessel, passed through the mercury and emerged. If the wire was fixed, the magnet turned around it: if the wire was mobile, it turned around the magnet. This was, in fact, the principle of the electromagnetic motor.

Ampère, to whom Faraday had sent a model of his equipment, perfected it. The Englishman Barlow made further improvements in 1822 by inventing a cog wheel of which the teeth alone dipped into the mercury and which turned in the field of a magnet when the current passed into these teeth.

The 'magnetism of rotation'

Faraday kept up-to-date with the experiments carried out by André Ampère with the help of his equipment. The only deductions he could make from them seemed confused, like that which claimed that electrical and magnetic attractions and repulsions were 'illusions'. In fact it seemed that what Faraday meant was that there was no fundamental difference between electricity and magnetism. His analysis of the two forces was being formed. Faraday was immersed in a flood of activity, overwhelming in its productivity: in 1823 he discovered the liquefaction of gases (see p. 117); in 1824 he discovered benzene and

---

An enigmatic forerunner

There was at least one scholar before Faraday who shared his intuition about force fields; this was the very mysterious Serbo-Croatian Jesuit, Ruggiero Giuseppe Boskovic, an astronomer and mathematician, who wrote in 1754 that matter consisted of particles joined by considerable forces capable of acting at a distance. In 1844 Faraday wrote prophetically: 'Particles are only the centres of forces; the force or forces are the constituent elements of matter. ... They are materially penetrable, probably to their very core.'

the polarization of light in a magnetic field; in 1825 he was elected to the Royal Institution. ... His work on electromagnetism was therefore rather intermittent, but he pursued it nevertheless.

In 1824 Faraday tried an idea that he had noted in 1822: to convert magnetism into electricity. He didn't succeed, and there were further failures in 1825 and 1828. Luck was not with Faraday during those years, but neither was it with Ampère, who in 1824 had come as close as possible to a major discovery in the course of an experiment. If a magnetized needle was suspended from a wire it oscillated; if a metal disc was placed underneath it, the oscillations disappeared almost completely. If the disc was turned, the needle turned in the same direction; and if the needle was turned, the disc, if it was mobile, also turned. This was very like Faraday's first discovery, but with this slight difference; in this case there was no electric current. A 20th-century observer would understand straight away that there has been induction of an electric current, but Ampère and his illustrious colleagues, Arago and Duhamel, spoke of 'magnetism of rotation' which is quite meaningless. In fact it was Ampère who discovered electromagnetic induction but since he understood nothing of it, he must be denied the honour.

Faraday, who also knew about these experiments, began to get a hazy understanding of their sense. He constructed new apparatus, again very simple. two conducting wires were wrapped separately round a ring of soft iron, which is a good conductor of electricity. The first wire was linked to a galvanometer, which measures current. The ends of the second wire were connected to an electric battery. If the current was flowing steadily, nothing at all happened; but when the current in the second wire was switched off or on, a current was briefly registered in the first wire, which had no contact at all with the second. This was a famous experiment; it showed that the variation of magnetic field caused by switching off or on a current in the second wire had in turn induced an electric current in the first wire.

Faraday repeated the experiment with different apparatus. He wrapped a conductive wire around a cylinder of soft iron.

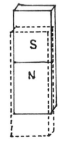

Rediscover it for yourself
One of the simplest and cheapest methods of verifying and rediscovering electromagnetic induction consists of making a coil of copper wire supplied with an electric light bulb and bringing a simple bar magnet close to it and then taking it away. This single movement is enough to create a current and light the bulb.

When he changed the magnetic polarity of the cylinder an electric current passed into the wire. In yet another simple variation, a magnet was introduced and withdrawn from a spiral coil of conductive wire, thus inducing an electric current in the wire each time.

Magneto and dynamo

The three experiments came one after the other in the space of a very few days on 24 September, 1 and 17 October 1831. On 28 October Faraday refined his apparatus: on to a spindle between the poles of a horseshoe magnet he fixed a copper disc which was joined to a galvanometer by two electric wires, one coming from the centre of the disc, the other from its outer edge. He made the wheel turn, and the galvanometer needle indicated the flow of a current. Faraday had just used electromagnetic induction to transform mechanical energy into electrical energy, thereby inventing the first prototype of a generating magneto with continuous current.

The scientific world threw itself passionately into developing this work: a few months later in 1832 the Italians Pixii and Dal Negro constructed, independently, generators with alternating current in which horseshoe magnets turned facing wire armature windings. In 1833 the Englishman Ritchie had the coil turn instead of the magnet. In 1834 the Russian Jacobi constructed an electric motor with electromagnets, which he would perfect in 1839 for the purpose of propelling a boat by electric motor. In 1866 the Englishman Wilde replaced the magnet with a battery electromagnet, and in 1867 the German Siemens stimulated the electromagnet by using the actual current of the machine: the magneto was transformed into a dynamo.

A laborious interpretation

Over and above the practical applications which were to continue to multiply, what has captured the attention of the scientific historians was the remarkable power of theoretical interpretation displayed by Faraday. What we have here is one of the great discoveries in the full sense of the word but its interpretation was laborious, although marked with the stamp of genius. In an early memo dated 1831, Faraday had the idea of a general law, 'very simple although difficult to express'; the principle of this law was that the phenomena depended on the way in which the conductor intercepted the magnetic field lines.

'The power to induce electric currents', he also wrote, 'is exerted by a magnetic resultant, or axis of force, exactly as circular magnetic actions are produced by an electric current.' Twenty years would have to pass before the genius of Maxwell, the great mathematical theoretician of electromagnetism, would exploit these notes fully in the form of his fundamental equations. It would seem that as far back as 1832 Faraday had a profound knowledge of the law of electromagnetic induction — and even, according to some historians, an inkling of the theory of fields and relativity. All that was lacking was a precise formulation of the law and especially of the electromotive force of induction. However, this pioneer was so far ahead of his time that the concepts with which he tried to surround his discovery would appear obscure, even when Maxwell had set them out in mathematical terms in 1855, then in 1861–2. The fact is that Maxwell and Faraday were both subject to intuitions which could only be verified by the physics of the 20th century, and which centred confusingly around atomic and particle dynamics. Their thoughts ran far ahead of words. As the French mathematician Henri Poincaré observed, Maxwell's mathematical definitions did not truly explain Faraday's discovery; they only showed that a mathematical explanation was possible. It was only from the 1880s onwards, thanks especially to the research of the German Hertz, that electromagnetic induction took its place in the general theories which led to relativity.

Embryonic star

Walker, Maloney, Wilking and Lada, 1986

Birth of a star

The gravitational collapse of a cloud
of gas and dust is now identified as
the first phase of the birth of a star.

Since the beginning of the 20th century and the birth of cosmology, thanks to the perfecting of optical telescopes and radio telescopes, astronomers have been discussing the method of formation of the stars. From the 1960s observation of the successive contractions which take place towards the end of the life of a star, going from the giant phase to that of a dwarf or a black hole (see p. 29) (obviously an indirect observation put together by deductions from astrophysics), has allowed for the development of a theoretical model: the star itself is formed by the gravitational collapse and heating up of a cloud of gas and dust.

Such clouds have been observed since the 1960s but then it was impossible to know what was happening at their core because they are opaque. They are in fact made up of very dense hydrogen and dust particles. From the 1970s the invention of infra-red astronomy and radio astronomy using millimetre waves has allowed for the structure of these clouds to be penetrated and for it to be shown that their centres concealed hot and active cores. Many infra red sources were then found in the interior of the gas and dust clouds; they exhibited all the anticipated and necessary characteristics corresponding to the definition of embryonic stars — also called protostars — but they also exhibited one other characteristic which disturbed astronomers: they violently expelled large quantities of matter into space and this phenomenon could not be reconciled with the gravitational collapse which governed their formation. In fact a star which is formed by the accretion of matter on to its core would have to be attracting and not expelling surrounding material.

IRAS 1629 A

In 1983 the American satellite IRAS (Infra-Red Astronomical Satellite) gathered in the course of its 10 months of operation, information on some 250 000 infra-red sources. One of these celestial bodies caught the attention of the team of astronomers at the University of Missouri — Walker, Maloney and Wilking, Lada, — who were studying the IRAS data; it was the source designated by the abbreviation IRAS 1629 A, in the cloud Rho Ophiuchi to the north of the star Antares in the constellation of the Scorpion, 520 light years (1 light year = 6 million million miles) from Earth. This object was very cold since its temperature was −233°C, that is only 40 degrees above absolute zero. However, its relative proximity made the study of it easier. Radio

---

Kant and Laplace

The discovery of the first embryonic star seemed to put an end to a controversy which began with Kant and Laplace, the first supposing that the original gaseous nebula condensed without any rotation, by simple contraction, the second claiming more correctly that there had to be rotation, although he was evidently unable to conceive of its mechanism.

study helped to improve our knowledge of it. Centred on the observation of carbon monosulphide (a molecule present in interstellar clouds which emits strong millimetre wave radiation when present at high densities), it revealed that in fact there was a dense core at the centre of IRAS 1629 A. The Doppler effect (see p. 56) indicated moreover that there was a movement of matter from the less dense exterior of the cloud towards the interior.

It was later revealed that the cloud, whose core measured some 600 thousand million km in diameter, was elliptical in shape and that it contained two jets of matter perpendicular to its main axis, a phenomenon already found in other infrared sources and now explained in the following way. The gravitational collapse can be visualized as the transformation of a football into a rugby ball under the influence of an internal force. This contraction, without altering the nature of the surrounding material, eventually leads to the expulsion of a certain fraction of its contents. The cloud is therefore subject to two distinct forces of opposite direction: one which forces gas and dust particles towards the core, the other which expels a certain fraction of these same gases and dust.

A comparison of the theoretical models and the available data on IRAS 1629 A indicates that its gravitational collapse began only about 30 000 years ago, which makes it an extremely young body in astronomical terms. In any case it would need only three times as many years, say a thousand centuries, according to Lada, for the central core, evidently much denser, to reach the diameter of the Sun, the com-

pression of its matter then having raised its temperature to levels sufficient to start nuclear reactions at its centre and power the newly formed star.

A star is born

Astronomers using the James Clerk Maxwell Telescope (JCMT) in Hawaii have discovered what seems to be the youngest star known. The protostar was first detected as a source of infrared radiation during the sky survey made by the Infrared Astronomy Satellite (IRAS) in 1983 in Nabula NGC 1333 — about 1100 light years from Earth. Astronomers Colin Aspin, Goeran Sandell and Bill Duncan of the Joint Astronomy Centre in Hawaii, Adrian Russell of the Royal Observatory, Edinburgh and the Max Planck Institute for Physics, Munich, and Ian Robson of Lancashire Polytechnic used the JCMT to study the region highlighted by IRAS; they discovered that the source — known as IRAS 4 — was surrounded by a shell of dust thicker than any yet seen around a young star. This almost certainly means that IRAS 4 is still forming, with dust from the surrounding shell falling onto it under the influence of gravity. The energy released during the impact of material landing on the protostar is the source of the object's heat. The centre of the protostar is not yet hot enough to have started the nuclear reactions which power fully developed stars. The amount of dust around IRAS 4 is the vital clue to its youth, implying that it is only a few thousand years old. Astronomers expect another million years to elapse before the star forming process is complete and IRAS 4 becomes visible as a normal star, although one several times more massive than our own Sun.

Endorphins

Guillemin, 1975; Hughes, 1975

A regulator of behaviour

These polypeptide molecules are in fact drugs secreted by the brain itself.

Since the 1960s numerous medical disciplines — anaesthesiology, neurology, physiology — have been interested in the behaviour of the nervous system in the presence of drugs. It is not in fact known how drugs and anaesthetics act on the brain. One theory was then current: specific receptors existed in the brain to pick up the opiate compounds. The verification of the existence of these receptors was a complex business: firstly because it had to be established that opiate substances do fix themselves at specific sites in the brain; secondly, and more difficult, it had to be established that the chemical affinity between the opiate molecule and the receptor is a close and equally specific affinity, in other words, that the receptor is exactly suited to the molecule. After much work the discovery of cerebral receptors was made, independently, by the Swede Terenius at Uppsala, and by the Americans Pert and Snyder on the one hand and Simon on the other. At the same time it was discovered that opiate receptors are not distributed just anywhere in the brain but are found in a particular area called the limbic system, also called 'the emotional brain'. To put it another way, the emotions might play a part in the fixing of opiates by the receptors.

A natural substance

This was an astonishing discovery, which was to lead quite logically to another. Given that opiate receptors had been found in the brains of animals not normally in contact with drugs — rats, pigs, monkeys, cattle — it was wondered what use these receptors were to them. In 1975 the American, Hughes, and the Frenchman, Guillemin (see p. 36), brought to the fore such substances in the brain of the pig. There were several described as imitating morphine, therefore morphomimetic, and which were at first called encephalins and then endorphins. It is not yet known if these are neurochemical substances of an entirely separate nature, that is to say produced just as they are by the nervous system, or if they are products of the weakening of the neurotransmitters. We can only make an imperfect guess as to their role.

A great many trials carried out since then indicate that endorphins play a role in the management of pain and also in numerous aspects of behaviour and physiology. They are more prolific in the first stages of emotional stress, they aid, or in the case of disequilibrium, inhibit, the process of learning, they have a direct action on muscle tone and they play a role in obesity.

Alcohol

One theory which is current but unproven is that the depletion of endorphins, resulting from stress, encourages the consumption of alcohol or drugs.

Jogging

Jogging enthusiasts having reported that in addition to a pain barrier they also experience a kind of intoxication which enables them to renew their efforts, neurophysiologists had the idea of measuring the products of the degradation of endorphins in sportsmen. They found abnormally high amounts, indicating that physical effort induces a more abundant secretion of endorphins. However, this might also be an early sign of cardiac lesion.

Enzymes

Payen, Persoz, 1833; Büchner, 1897

Vital intermediaries

These important proteins were only identified in the last century. Without them there would be no alcohol — nor indeed life.

In ancient times the process of fermentation was not understood, and the transformation of vegetable or plant juices and saps was attributed to gods like Bacchus. The responsible agents were in fact enzymes which also cause fermentations other than those of sugars (eg curdled milk).

Fermentation

More and more interest was shown in the industrial fermentation of alcohol because it was economically important. In 1814 the German Kirchhoff had maintained that in the manufacture of beer the germinating barley or malt acted as a catalyst, transforming the starch into glucose. The phenomenon interested two French chemists, Anselme Payen, discoverer of cellulose, and Jean-François Persoz, who discovered that it was not the malt but a substance secreted by it which triggered the formation of dextrin and sugar from the starch. That was in 1833. They named this substance diastase, a name which was to be replaced during the 20th century by 'enzyme'. In 1837 the German Liebig discovered another enzyme, emulsin, and throughout the century other scientists were to discover more: lipase (Claude Bernard, 1849, and Pelouze, 1855), saccharase (Berthelot), laccase and tyrosinase (Bertrand). The true discoverers of enzymes were therefore Payen and Persoz, but their discovery was long neglected and when Pasteur analysed alcoholic fermentation in 1857 he took hardly any account of it. Payen and Persoz had indicated that the fermentation of malt could occur without any living input; and Berthelot had reinforced this hypothesis by demonstrating that a saliva enzyme that he had isolated, invertin, automatically transformed sugar solution into simpler sugars. It was not until 1897 that the German Eduard Büchner managed to isolate diastase and to demonstrate at the same time that the intuition of Payen, Persoz and Berthelot was well founded: diastase keeps its effectiveness even when isolated from yeast.

Interest in enzymes, about a thousand of which have now been discovered, goes far beyond the processes of alcoholic fermentation and bread-making. They are essential to a large number of biological processes; the whole metabolic balance of living beings depends on the intervention of enzymes. It has been demonstrated that enzyme deficiencies are all of genetic origin.

Proteins which effect 'transmutation'

Enzymes are omnipresent proteins: all the cells of mammals contain them, and any one of our cells has about 3000 of them. They facilitate chemical reactions within the limits of temperatures compatible with human life; without enzymes either the reactions in question would not occur or they would occur at temperatures approaching 300°C.

'Transmutations'

It has been suggested that enzymes bring about transmutations; the term is too extravagant, for it only applies to phenomena in which there has been transformation of the cell nucleus. However, the power of enzymes is astonishing: a single molecule of an enzyme can produce one million million molecules of oxygen per second!

The extra chromosome

Turpin, Gautier and Lejeune, 1958

Down's syndrome

*The identification under the
microscope of the cause of
mongolism or Down's syndrome
helped in the understanding of many
other genetic illnesses.*

For centuries people have been born who are not only mentally retarded but also exhibit particular physiological characteristics, their features have an Asiatic cast, mainly because of an oblique fold of skin over the eyelid and inner angle of the eye specific to Mongolian peoples (the epicanthus). The eyes are very widely spaced; the neck is short and flat at the nape; the limbs are small but the hands and feet are large; and the lines of the hand (dermatoglyphics) are few and atypical, most often consisting of one isolated palmar crease. Muscle tone is weak and the individual is vulnerable to all sorts of infections. The child may reach maturity but in most cases serious malformation of the organs, such as anal perforation, heart defects or stenosis (narrowing) of the digestive system shorten an existence which, although not unhappy, is somewhat limited. The person with Down's syndrome whose average mental age is no more than six years, has a happy and affectionate nature.

These are the principal characteristics described for the first time by the Frenchman Seguin in 1844. For several decades a 'genetic defect' would be suspected. This vague expression implied that the disease was congenital without being able to state whether or not it was hereditary nor why a person with Down's syndrome could have brothers and sisters who were completely normal.

One chromosome too many

The existence and structures of chromosomes had been known since 1920: they are made up of rows of independent genes. It was not until 1958, however, that the Frenchmen Turpin, Gautier and Lejeune thought of testing the idea that the cause of Down's syndrome might be a genetic disorder. This could in fact be verified quite simply with an optical microscope. (The electronic microscope would only prove to be of real use for the examination of the gene in 1969.) It was necessary, however, to separate the chromosomes in such a way that they could be observed distinctly; this is normally impossible, for they form a confused tangle. The scheme thought up by the French researchers consisted of separating the chromosomes by plunging them into a hypotonic solution.

They discovered that cells from a person affected by Down's syndrome do not have 46 chromosomes — that is, 23 pairs — but

> **Theory**
> Sociobiology has postulated that chromosomes carry not only physical and physiological characteristics but also psychological traits. However, this theory of the hereditary nature of intelligence, as it were, is far from unanimously held.

47. The extra chromosome is found in the place occupied by the normal chromosome 21; it is termed trisomy 21. The formula of this triple chromosome can be XXY or XYY. It would be established later that the extra chromosome can escape notice when it is not free but has attached itself to another type of chromosome after being displaced or 'translocated'.

In certain cases the presence of an extra chromosome may not give rise to clinical signs, precisely because its final position inhibits their expression; this is called a balancing translocation.

Other related syndromes

To this fundamental discovery would be added those of Klinefelter's syndrome, where the extra chromosome is attached to the sex chromosome and where several extra chromosomes may even occur giving formulae as aberrant as XXXXY, XXYY, XXXYY, or XXXY; Edwards syndrome, or trisomy 18; Patau's syndrome, or trisomy 13; Turner's syndrome, which is in fact an unpaired sex chromosome; Wolf's syndrome, characterized by the absence of the short arm of chromosome 4; and *cri du chat* syndrome, so-called because children suffering from it have a piercing cry resembling that of a cat, and characterized by the omission of part of the short arm of chromosome 5.

So opened a whole new chapter in genetics which would make for a better understanding of hitherto obscure diseases. Meanwhile the term 'mongolism' is gradually being replaced, at least in the medical world, by that of trisomy 21 or Down's syndrome. The latter term pays homage to the 19th-century doctor who elaborated on the description made earlier by Seguin.

The ageing of the ovum

Trisomy 21 or Down's syndrome is not congenital, save in exceptional cases; it manifests itself particularly in pregnancies of women over 40 whose ova are impaired as a result of the ageing process. In 1962 an Australian study showed that the risk of Down's syndrome increased considerably with the age of the mother. In mothers under 20 the risk is 1 in 2370 births; for those aged between 20 and 24, 1 in 1600; between 25 and 29, 1 in 1200; and between 30 and 34, 1 in 870. For those between 35 and 39 the risk increases markedly, being 1 in 300; in the age group 40 to 44 it is 1 in 100; and in the case of mothers over 45, 1 child in 46 has Down's syndrome.

Chemical causes

In addition to the risks inherent in the advanced age of the mother, there emerged those associated with viral diseases, certain chemical products, including alcohol and tobacco, and radiation. All these are capable of triggering deletions or translocations in the chromosomes with more or less classical results for the embryo: malformations of the central nervous system with mental retardation, microcephaly (an abnormally small head), anencephaly (failure of the brain to develop) in certain cases, and various malformations of the organs. A similar picture emerged in the 1980s with the discovery of oncogenes, dormant genes fixed on various chromosomes whose awakening or activation by chemical products or ionizing rays unleashes the spectre of cancers, cardiovascular and hormonal diseases, and immunity- and enzyme-related disorders.

Chromosome of crime?

In 1966 it was discovered that Richard Speck, killer of eight Chicago nurses, carried an extra Y chromosome. Two years later another killer, Daniel Hugan, was found to have the same extra Y chromosome. Professors Turpin, Gautier and Lejeune examined him and attested to loss of criminal responsibility. Moreover, 7 out of 196 prisoners in the State Hospital at Carstairs, Scotland, were found to have trisomy XYY. The expert Linus Pauling explained that chromosome Y could disturb the regular irrigation of the brain and so lead to criminal acts. The expression 'chromosome of crime' was born. It was to go out of fashion when it became clear that there were also cases where carriers of the extra chromosome exhibited no psychological abnormalities whatsoever.

Fermentation

Cagniard de la Tour, 1835; Schwann, Müller and Kützing, 1835;
Pasteur, 1854

The key to putrefaction

For centuries the decomposition of organic matter into new molecules remained a total mystery.

Fermentation is the decomposition of organic substances induced by micro-organisms or enzymes, leading to the formation of new molecules. It remained totally unexplained until the 19th century. In ancient times it was attributed to the gods or to demons.

Its discovery is generally attributed to Pasteur, who would have been the first to observe with a microscope the greyish deposit in the containers where a faulty alcoholic fermentation had been produced. In reality globules of the brewer's yeast were observed for the first time in 1680 by the ingenious inventor of the microscope, the Dutchman Antonie Van Leeuwenhoek (see p. 102), but he had not specifically explained fermentation. The discovery was made again in 1835 by the French physicist Cagniard de la Tour, who also observed globules of brewer's yeast under a microscope and their multiplication by 'budding'; he concluded quite correctly that yeast is 'an organism whose vegetation most probably leads to a release of carbonic acid and alcohol'. The Germans Schwann,

Müller and Kützing made the same observation in the same year. Not only did no-one take any notice; it was in fact rejected in favour of the interpretations of two much more prestigious chemists, Berzelius and Liebig, which were completely false! According to Liebig fermentation proceeded from spontaneous decomposition; according to Berzelius it was a reaction to contact.

Pasteurization

It was these ideas that Pasteur came up against in 1854 when, begged by industrialists to find a way round the irregularities of alcoholic fermentation, he observed the greyish deposits, under a microscope and rediscovered the fermenting agents for a third time. Pasteur's genius consisted in going beyond simple observation: he extracted a trace of the deposits and added it to a liquor made of sugar and crystallized mineral salts. The substance, a yeast, proliferated by using the sugar; however, it did not produce alcohol, as in normal fermentation, but lactic acid. Pasteur had discovered two facts: that fermentation is caused by yeasts, and that defective fermentation is caused by a parasitic yeast. He also noted that parasitic yeasts develop after initial alcoholic fermentation and concluded that they can be killed by heating the wine or beer to 55°C, which does not alter its taste: pasteurization had begun.

Pasteur explains

In order to tell the producers about the benefits of his discovery, Pasteur himself went round the vinegar makers, wine makers and brewers to explain how to ensure its best application: by the selection of appropriate yeasts and the destruction of others by heat, by random inspection of the vats, and by the protection of musts against impurities.

Numerous applications

Pasteurization spread rapidly: all the distilleries benefitted from it, as well as the food industry, since milk and other drinks could also be preserved in this way. Once on his way Pasteur would discover anaerobic fermentation, which can only occur in the absence of oxygen.'

Fertilization

Hertwig, 1875

How offspring are born

Until the last quarter of the 19th century it was not known that the egg and sperm fused into a single nucleus.

The observations of the layman as much as the experiments of the scientist had shown for centuries that sexual union between a male and a female was necessary for the production of another individual and the perpetuation of the species, but there was no scientific basis for this empirical knowledge until the end of the 19th century. One day in 1875, one of the two Hertwig brothers, Oskar, professor of anatomy at Jena University, was observing the egg of a sea urchin against the light and noted that the nucleus of the spermatozoon and that of the ovum had just merged into one. He was the first in the history of the sciences to identify the phenomenon of fertilization.

It was not that the phenomenon had not interested scholars. In 1827 the German von Baer had discovered the mammalian ovum in a bitch. Three years before that the Frenchmen Prévost and Dumas had even experimentally fertilized some frog's eggs, noting that the filtered male liquor lost its fertilizing power while the residue retained it. So the idea of fertilization existed well before Hertwig's discovery, but it was confused, for spermatozoa were not known (see p. 193) until the work of Kolliker, Wagner and Leucart (1841 and 1849). It was therefore about a quarter of a century before Hertwig's discovery that any clear idea emerged of the nature of spermatozoa: true cells.

Other researchers observed the penetration of the ovum by the sperm in the sea urchin and then in the starfish. The next important step was taken by the German, Walther Flemming, who, in 1882, observed the fusion of gametes (sperm and ova) at the moment of fertilization, then the division or meiosis of the first two cells, each of which reconstituted the genetic capital of the egg.

Embryology and histology

Since then other discoveries have followed, leading to the science of embryology. This would be strongly influenced by von Baer's doctrine: the embryo is formed by leaves or layers of cells (ectoderm, mesoderm and endoderm), an idea which was itself to give birth to histology or the study of the structure of tissues. Until the beginning of the 20th century embryology could only be applied to animals. The youngest human embryo that had been studied was 14 days old, and it was not until 1926 that the American, Streeter, was able to study a younger one, aged 11 days. Not until the 1940s would we begin to have a fairly precise idea of the first days of the embryo and the implantation of the fertilized ovum.

The lingering darkness
In the 19th century very many 'scholars' believed in spontaneous generation — Pouchet, for example, Pasteur's unfortunate adversary. It was then taught in the universities that mice could be 'spontaneously' born from piles of dirty linen. The rare intellectual discipline of a Hertwig or a Pasteur was needed to observe fundamental biological phenomena with an objective attitude.

Fingerprints

Purkinje, 1823

The incomparable signature

Fingerprints were in use as far back as the Babylonians. Their uniqueness, however, was not established until the 19th century.

The kings of Babylon who wanted to give their edicts the stamp of incontestable authenticity imposed the print of their right hand in the clay below the engraved text before it was fired. The Babylonians, like numerous other peoples of antiquity, knew that there are no two hands whose prints are identical.

The practice of using fingerprints lasted for centuries, as long as few people knew how to sign their name. It was only in 1823 that the Czech Jan Evangelista Purkinje, founder of experimental physiology, who was studying sweat glands, discovered that there are no two individuals in whom the patterns of ridges and grooves of the skin are identical. He came across this fact because the sweat glands open out into the depressions of the grooves.

It was only then that science officially acknowledged the individuality of fingerprints; but it would take another half-century before the taking of fingerprints in ink (dactyloscopy) would pass into the realm of judicial anthropology. (Anthropometry is the study of sizes and proportions of the human body.) The first known user of prints taken in ink, or dactylograms, was an Argentinian police officer from Buenos Aires, Juan Vucetich, who published his treatise on comparative dactyloscopy in 1888.

In the 1890s the British developed a comparable system, called the Galton Henry system, which was put into operation by Scotland Yard in June 1900. Three years later Alphonse Bertillon, founder of French

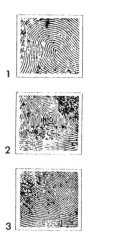

Three basic patterns
1. *the loop*
2. *the whorl*
3. *the arc*

criminal anthropometry, in turn used dactyloscopy. Nowadays it is used by all the judicial identification departments in the world.

Arcs, whorls and loops

Three main patterns have been identified in fingerprints: the arc, the whorl and the loop. Comparative anthropology has established that their frequency varies according to ethnic origin. Europeans have few arcs, Pygmies and Bushmen of Africa have the most (10–16 per cent); and Orientals have more whorls than Occidentals (a maximum of 42 per cent in the former and 16 per cent in the latter). Loops are the most frequent, Europeans, Africans and the Ainus of Japan having the most (52–76 per cent of the population) while the rest of the population has far fewer (28–64 per cent of individuals). On the whole Occidentals have more arcs and loops than Orientals. The division of the population according to the distribution of these frequencies corresponds in many cases to divisions made according to blood groups.

Galvanotherapy

Galvani, 1786

An ephemeral therapy

Not until a century after its discovery was the action of a continuous current on an organism taken seriously.

The terms 'galvanotherapy' and 'galvanism' are hardly used today, but their principle and its applications continue to be developed.

In 1786 the anatomist and doctor Luigi Galvani, of Bologna, discovered that the sparks produced by one of the pieces of electrical equipment in his laboratory caused muscular contractions in a dissected frog. Other experiments, including one in which a nerve was connected to a muscle by an electric wire, convinced him of the existence of an electric current in living tissues. The first application of this famous discovery was the treatment of a case of facial paralysis by continuous current, carried out in 1795 by the Frenchman Halle. Doctors like Magendie and scientists like Faraday took the discovery further, thereby prefiguring electric diagnosis by both electrocardiogram and electroencephalogram (see p. 59). Electrotherapy, born out of galvanotherapy, has applications of proven worth: the cardiac pacemaker for instance, and the treatment of bone fractures.

In patients with heart block (a malfunction of the sino-atrial node or conducting system) artificial pacemakers (usually battery operated and implanted under the skin) stimulate the heart electrically to restore and maintain higher rates of cardiac contraction.

D'Arsonvalization

In the first decades of the 20th century, the Frenchman d'Arsonval and his followers believed they could treat a great many illnesses by application of an electric current, or d'arsonvalization. The results were hardly convincing and this form of therapy disappeared after the 1939–45 war.

Gas lighting

Lebon, 1799

Let there be light

Using coal gas for lighting was attempting the impossible.

Before 1780 there was no such thing as organic chemistry and practically nothing was known about organic compounds such as those of plants and of wood. Mineral chemistry had just been born, for it was only in 1772 that nitrogen had been discovered (see p. 139), and the phenomenon of combustion remained relatively obscure, for the experts were bogged down in theories of 'phlogistics'.

However, in 1785, aged barely 18, the future engineer and chemist Philippe Lebon had the idea of distilling wood or charcoal in order to extract a gas which he thought could be used to light streets and houses.

'Thermolamps'

His researches lasted several years and were conclusive enough for him to begin practical trials in 1797. On 28 September 1799 he took out a patent on the use of wood gas or charcoal gas for the purpose of producing light and heat, with the help of devices which he called 'thermolamps'. His invention was to remain theoretical for a long time, but it had achieved enough renown for Lebon to be invited to develop it as the means of lighting the imperial coronation celebrations in 1804. That was the year in which Lebon was assassinated in mysterious circumstances at the age of 35.

At around the same time the Irishman Murdoch and the Englishman Winsor (real name Winzler) had had the same idea; in the year after Lebon's death Winsor succeeded in installing the first (limited) system of gas lighting. This gas was in fact a mixture of carbon monoxide (about 33 per cent), and nitrogen (about 65 per cent), the remainder being hydrogen. It was evidently the hydrogen which burned. Paris was not to benefit from 'Lebon gas' until 1829. Meanwhile, however, the butterfly burner, which spread the flame in a fan shape and gave it more luminosity and more lighting power, had been invented. Two further refinements would follow: holophane lighting, which consisted of surrounding the burner with a spherical opalized glass shade to assure a homogenous diffusion of light, was invented by the Frenchman Blondel in 1893; and the Auer burner, which enhanced the luminosity of the flame by having it pass through a mantle of thorium oxide, was invented in 1895 by an Austrian of the same name.

The spread of electric light, which had begun with the invention of the incandescent lamp in 1878 (Edison), was to bring about the progressive decline in gas lighting. Yet the genius of Lebon remains astonishing. Motivated only by intuition he not only completed an invention far in advance of the chemistry of his time, but also imagined an urban system of lighting and heating, the principle of which persists to the present day, with obvious amendments like the replacement of wood gas by natural gas for heating and other domestic uses.

A dazzling genius

Philippe Lebon was one of the greatest geniuses of the history of science. The notes which he left indicate that he had conceived of an engine, quite revolutionary for his times since it would have operated on gas, with an electric injection pump and an electric spark ignition. ... He was also the first to envisage, in 1801, an improvement in the output of such an engine thanks to the compression of gases.

Gene amplification

Mouchès, Bergé, Silvestri, Pasteur, Raymond, Hyrien, de Saint-Vincent, Gheorghiou, 1986

How insects resist insecticides

The progressive adaptation of insects to insecticides has resulted from a massive genetic modification.

Scientists undertaking the struggle against insect pests have met with a series of nasty surprises ever since systematic investigations began. This was around the 1920s, when there was increasing use in agriculture of dichlorodiphenyltrichloroethane, discovered in Europe in 1870 and better known as DDT.

Heightened resistance

The first and most general discovery made was that DDT and the other insecticides and pesticides harmed not only those insects defined as 'pests' — a description which has broadened considerably since the 1950s — but was also equally harmful to other insects and to numerous species of animal. This is how DDT has come to be found in the liver of seals in the Arctic regions. Human beings themselves were found to be suffering from the excessive use of DDT and other similar (chlorine-based) products since DDT was found in the milk of nursing mothers.

The second discovery was more specific but no less important. After a few years during which a decrease in the population of species of pest insects was recorded due to the introduction of new chlorine-based pesticides, a definite resistance to these products was noted. In particular, since the 1960s we have witnessed the appearance of mosquitoes, as well as locusts and other insects, on which the chlorine-based pesticides no longer have any effect. These cases of resistance multiplied in the 1970s, and the time between the commercial introduction of a product and the appearance of colonies of resistant insects was observed to grow less. In 1986, for example, cases of resistance were recorded in respect of a group of insecticides only in commercial use since 1980.

Questions

The phenomenon had given rise to entomological and genetic studies from the 1960s onwards but no satisfactory explanation had been found. The most widely held theory was that the insecticides decimated the greater part of any given insect population but spared a tiny percentage of resistant individuals from which the species regained its previous demographic density. This theory was intellectually attractive; moreover, it was exactly parallel to the one which had explained in similar terms the appearance of antibiotic-resistant bacteria (see p. 80), but it did leave several points unexplained. The most significant of these were as follows: what would characterize the hypothetical resistance of certain individuals to chemical products? Whatever that resistance might be, how could we explain the inefficacy of chemical products whose power had already been demonstrated?

A genetic mutation

In 1985 a team of researchers, Mouchès, Bergé, Silvestri, Pasteur, Raymond, Hyrien, de Saint-Vincent and Gheorghiou, undertook to study the genetic inheritance of those colonies which had become resistant, and to compare it with that of colonies which had not; the example studied was a mosquito originating in California.

A considerable surprise awaited the researchers: a certain gene in the mosquito's genetic inheritance, present as a single specimen in the non-resistant strain, occurred some 200 times in the resistant strain. This was an unprecedented discovery in the history of genetics. The gene in question is one which enables the insect to synthesize an enzyme which destroys the insecticide. As a single gene was not sufficient to allow the insect to triumph over the insecticide it was repeatedly copied until the organism became capable of secreting enough enzymes to destroy the insecticide. It is possible that the multiplication of genes corresponds to the amount of insecticide present in the environment in a given period. In this way the Californian mosquito was able to resist the insecticide in doses much higher than what was thought to be fatal.

The phenomenon had also been observed in plants, then in cultures of cancerous cells. It is known by the term 'genetic amplification' and is of considerable practical and theoretical import.

A new strategy

Genetic amplification was first applied in the field of agronomics. It was evident that the fight against pests as it had been pursued since 1920 was mistaken, indeed harmful, because it created resistant strains. It is likely that we will see its replacement by a form of biological attack which consists of combating the insects with natural enemies: other predators, for example, or perhaps certain bacteria. It will certainly revolutionize the perfecting of new insecticides, which will now have to be tested from the point of view of any genetic mutations they might unleash. We must, in the near future, find chemical substances to which the insects will have no genetic answer. Those researchers mentioned above and attached to the Institute of Agronomic Research plan to modify the genetic inheritance of useful species to make them able to withstand the insecticides which are killing them because they lack the resources of genetic adaptation.

From the human point of view genetic amplification opens up new possibilities in cancer research: it allows for a better understanding of why cancerous cells are able to resist the assaults of the immune defence system and of chemotherapy.

Major implications

It is, however, in the field of general biology that the discovery of genetic amplification is most useful. In the first place it reveals the existence of a previously unknown mechanism of adaptation. Until this point it had been agreed that the karyotype (the appearance, number and arrangement of chromosomes in a cell) lent itself to certain variations and that it could draw on reserves of redundant sequences of DNA, intervening sequences or introns, in order to correct deficiencies or to bring about genetic recovery. The 200-fold reproduction of the resistant gene in the Californian mosquito indicates, at first sight, that there exists a mechanism of regulation much more powerful and far-reaching than recourse to introns. It remains to be seen how this mechanism is governed (doubtless by other genes), where it is situated, and whether it is present in all living species or simply in invertebrates. (Its discovery in cancerous cells leads us to suppose that it also exists in higher vertebrates including man.) The Frenchman, Ruffié, believed that a gene corresponds to the necessity for synthesis of a peptide (sequence of amino acids) itself situated in molecules (ie DNA) which are indispensable in the perfecting of a character. This thesis echoes the

'Harmful' and 'useful'

The old distinction between 'harmful' insects and 'useful' insects has been considerably weakened in the second half of the 20th century. It is now believed that any species can become harmful above a certain level of population and that equally it may become useful below that level. Both excess and insufficiency may bring about ecological imbalances. Exception is, of course, made for certain species like the mosquito, a carrier of viruses and parasites whatever its population level.

American theory of the 'selfish gene'; in some way DNA produces itself only in order to be able to continue reproducing. So we can conclude that among the mosquitoes of California the genetic inheritance has reacted to the attack of an insecticide, which blocked one or several indispensable genes, by multiplying this same gene several dozen times. The mechanism has therefore gone beyond the simple nature of contingent adaptation; it has become a major biological mechanism.

Transmissible and irreversible?

In the second place, the mechanism of genetic amplification immediately calls into question the profound nature of adaptation of species and the transmission of acquired characteristics (see p. 90). It seems certain that (in the Californian mosquito) genetic amplification is transmissible. If it is reversible, that is to say if the mosquito reverts to its old karyotype after removal of insecticides from its environment, then there is no essential mutation of species, that is to say no appearance of a new species. If the contrary is true there will have been a modification of the gene pattern, the resistant mosquito no longer being genetically the same as it was. We can now understand the importance of the discovery of gene amplification. Its general interest goes far beyond its specific import, which is in itself considerable. Insects are the carriers of many serious illnesses (see p. 104) like malaria, yellow fever and onchocerciasis. The control of insect populations governs world health, particularly in underdeveloped tropical regions where they do most damage, but in developed countries too.

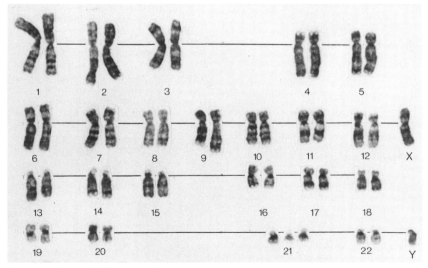

Karyotype of a boy with Down's syndrome

Genes (Chemical regulation by)

Monod, Cohn, 1952; Jacob, 1956

Metabolism is linked to heredity

*Complex equilibriums governed by
genes regulate the whole
metabolism.*

Every living being, from bacteria to humans, only lives thanks to the assimilation of nutritious elements available externally and incorporated internally by the mediation of enzymes (see p. 66). Since the beginning of the 1940s it has been accepted by biologists that enzymes, key elements in assimilation, are under the influence of genes, which decide whether or not they are to be secreted. It was thought that there existed in the cell a kind of 'government' which took its decisions according to the information it received, a sort of automatic control mechanism whose function was to assure survival of the cell by means of constant adaptation.

Enzymatic adaptation

This theory was correct but it was founded on an assumption whose principal defender was the American biologist, Hinshelwood: according to him the relationship of the environment with the cell would be direct and constitute the main element in the cell's regulation. Thus, the environment would be bringing new conditions to the cell which in some way the cell would 'know' how to handle, according to a definitive blueprint called 'teleonomy', the cell knowing not only what it must do but also how it should 'end up'. Attempts were made to explain general evolution in this way, and also cellular differentiation in superior organisms.

In 1947 the Frenchman Jacques Monod, studying this outline in the ordinary bacterium *Escherichia coli*, a favourite organism of biologists, made a first 'minor' discovery: when different saccharides (sugars), like glucose and lactose, were given to a culture of *E. coli*, it was the glucose which was immediately consumed, the lactose not being touched. The culture grew normally then stopped after a few hours, once the glucose had been consumed. Then it turned its attention to the lactose. For some time the phenomenon remained a mystery. Then it was noted that the enzyme which allows the lactose to be consumed, the lactase or β-galactosidase, only appeared when the last trace of glucose in the nutritional medium had been consumed. It was that enzyme which allowed the lactose to be broken down into its component parts: glucose and galactose. With his particular genius, Monod characterized the two stages of the phenomenon: the consumption of

---

A new outlook on physiology

Any organism, from a bacterium to a human being, assimilates its food and secretes its enzymes (and in the case of man its hormones) only under genetic direction. At a stroke, the whole of physiology was seen in a new light. The discoveries of Monod, Cohn and Jacob were to overturn the very notions of life and the mechanics of evolution.

glucose is 'constitutive', that of lactose, 'adaptive'.

Understanding of the significance of this laboratory curiosity dawned and with it the need to qualify Hinshelwood's theory. Contrary to what had been believed, the cell does not express the whole of its genetic possibilities in every circumstance; all genes are not active at the same time, nor are all the enzymes produced. The environment can therefore inhibit certain genes and, in this case, it does seem that there is inhibition of the gene governing the synthesis of β-galactosidase. The idea of Hinshelwood, a Nobel prizewinner whose authority was considerable, that the production of an enzyme is preceded by that of a precursor or pre-enzyme induced by the environment, needed therefore to be heavily revised. So why in fact did the forerunner of the β-galactosidase achieve synthesis of this enzyme only after the disappearance of the glucose?

'Pyjama'

It was then that Monod and the American Cohn made a vital discovery which was to lead to a whole train of deductions. The β-galactosidase could be secreted even when the sugar which followed the glucose was unusable! Hence the sensational collapse of the theory of enzymatic adaptation. The appearance of this enzyme is not linked to the consumption of lactose; it is simply produced once there is no longer any glucose. So the cell does not always 'know' what it is doing in as precise a manner as claimed in Hinshelwood's telenomy theory.

In 1956 Monod and his colleague François Jacob, with whom he was to share a Nobel prize, discovered the existence of two types of genes. In their famous experiment, humorously called 'pyjama' (because it consisted of 'dressing' one bacterium with the properties of another), they discovered that there existed, on the one hand, structural genes which code the production of proteins, including enzymes, and, on the other hand, regulatory genes which activate or inhibit the function of the first type. These two types of genes are arranged in sets called operons and the operons are interdependent. The discovery of the secretion of β-galactosidase is explained by the fact that the presence of glucose inhibits the production of this enzyme; this inhibition is only lifted with the disappearance of the glucose. So there are repressor genes (isolated by the American Gilbert in 1966) and activator genes (see p. 78).

Absolute power

Several years later, it would again be discovered that the mechanism is more complex and that it is not sufficient to inactivate the repressor of the lactose operon for it to begin secretion of β-galactosidase: it is also necessary for the cellular level of a substance called cyclic AMP (adenosine monophosphate) to have fallen to the level which triggers the production of the enzyme.

Cells appear then to be mechanisms endowed with very delicate and precise regulating systems, which govern the ability of the tissues which they constitute to maintain their function in different conditions.

Genes and evolution

The regulation of chemical processes by genes implies that every organism and its environment are closely interdependent and constantly exchange information. It also tends to indicate that the evolution of the species could not have occurred under genetic influence alone, nor under environmental influence alone but, probably, always by accident: be it a change in environment, or be it a genetic modification either spontaneous or forced by the necessity of adapting to the new environment.

Genetic recombination

Griffith, 1928

Genes have their reasons

The replacement, displacement,
suppression or addition of elements
of DNA can entirely modify the
nature of an organism, either
primary or evolved.

Genetic recombination is the modification of the genetic material of the DNA of any living being (see p. 52), from the top of the scale to the bottom, that is to say from man to bacteria. This modification may concern only a tiny fraction of the 100 000 or so genes or the 50 000 pairs of alleles in a human being, for example, and yet it assumes considerable importance, since it can modify one or several fundamental and hereditary traits of its nature.

This modification can operate spontaneously and then induce mutations which are also spontaneous; or it may be provoked by means now at the disposal of genetic engineers. In the latter case, it is called artificial (see p. 127 and p. 101). Although

> **Allele**
> An abbreviation for allelomorph. Any one of the alternative forms of a specified gene. Any gene may have several different types.

closely linked to mutations, since it causes them, genetic recombination is a phenomenon which is independent and specific, the considerable significance of which is quite different from that of mutation.

The pneumococcus capsule

Genetic recombination was discovered by bacteriologists. In the 1920s bacteriology was making considerable progress and studies were being made of one of the most dangerous bacteria, the pneumococcus, a spherical germ which is responsible for pneumonia. Several types of pneumococcus exist and it was learnt that they could be identified in the following way. If animals were inoculated with it, the pneumococcus secreted a protective capsule, a kind of viscous envelope to protect itself from the antibodies of the immune system. As this capsule varied in composition according to the type of germ, it was therefore possible to build up a list of the different sorts. This list was significant in that it opened up pharmaceutical possibilities, the composition of the capsule indicating the route

> **Two enigmas**
> Still unfound is the answer to one of the questions posed by Griffith's discovery in 1928: how does the transfer of DNA from the dead bacteria S to the living bacteria R come about? The passing of genetic material from deceased bacteria to living ones remains entirely enigmatic. Equally puzzling is the fact that the transfer of genetic material by the plasmids is made not only within a single species — for example, from one *Escherichia coli* to another — but also from one species to another, as Griffith demonstrated to everyone's amazement.

which must be followed in order to destroy it and thus render the pneumococcus vulnerable to the immune defences of the organism which it had attacked. Drawn up from pneumococci cultivated in vitro (ie in the laboratory) this list revealed a peculiarity often noted in micro-organisms cultivated in this way: the descendants of certain pneumococci lost the capacity to produce the protective capsule.

The first recombination

Studying these capsules, Dr Fred Griffith of the British Ministry of Health, noted that pneumococci with a rough capsule, which he designated by the letter R, did not make mice ill when he infected them, while the bacteria with a smooth capsule, which he designated by the letter S, killed them. He had R bacteria incubated with S bacteria (the latter first killed by heat), and inoculated mice with them. Several of these mice died.

This historic experiment constitutes the first proof of natural genetic recombination (although neither Griffith nor the other biologists of the time could understand the meaning of it, since DNA had not yet been discovered). Only the crude result of Griffith's experiment could be considered and this threw the scientific community into disarray. As is the practice when the results of experiments are very strange, numerous biologists repeated the work which had led to Griffith's discovery, using every possible precaution: the results were identical to those Griffith had found.

Missing out on the Nobel prize

One of the most astonishing omissions of the Nobel jury concerned a major discovery in biology: that made by Avery, McLeod and MacCarthy when they established, in 1944, that it was indeed DNA which was passing from the aggressive S bacteria to the harmless R bacteria, thus modifying the behaviour of the latter bacteria. The omission is perhaps explained by the unconfessed incredulity of scholars with regard to the discovery. The very idea of the transmission of genetic information from a dead organism to another left biologists sceptical, not to say scandalized.

Conquering scepticism

It took some 10 years for the general — and uneasy — scepticism which had greeted Griffith's discovery to dissipate. Biologists were forced to admit that the dead S bacteria communicated to the living R bacteria an element which modified the elementary characteristics of the latter; if it was not a gene proper, it was something very close to it. In 1944 the Americans Avery, McLeod and MacCarthy concluded in an important report that only the transfer of DNA from the S bacteria to the R bacteria could explain Griffith's discovery.

There was no reason to think that the transfer of genetic material only occurred from one type of pneumococcus to another; there must also be other similar transfers among other species of bacteria, changing harmless bacteria into pathogenic or disease-causing ones.

The plasmid

The actual mechanism of transfer remained obscure for many years despite the considerable amount of work carried out in the United States to clarify it. The first part of the explanation was provided in 1955 by the case of a Japanese woman who had contracted incurable dysentery caused by a type of *Shigella* bacterium. The germ responsible possessed the unheard-of property of being resistant to the sulphonamides, to streptomycin, to chloramphenicol and to the tetracyclines. Such resistance to antibiotics sparked off research in Japan into several areas. In 1960 the Japanese bacteriologists discovered the key to the enigma: the *Shigella* bacterium had acquired its resistance from an ordinary intestinal bacterium, *Escherichia coli* and had then transmitted it to other *E. coli* and to other *Shigellae*. This ability to resist was not chromosomal, so a genetic name, episome, was found for it, indicating a vague entity. In reality the episome had been discovered in 1952 by the American Lederberg; it was a bacterial cellular element independent of the chromosomes, but including a small quantity of genetic material, the plasmid. However, neither the Japanese, nor the biologists of other countries made any connection at first between the mysterious episome which had been at work in the intestines of the Japanese woman and the plasmid. The connection was not finally made until 1965. An inventory was then begun, not yet finished, of all the varieties of bacterial plasmids.

The genetic recombination which made certain germs resistant to antibiotics was brought about through the transfer of plasmids containing genetic information, a few meagre strands of DNA, which allowed the bacteria to secrete enzymes which could destroy antibiotics; the transfer between resistant and non-resistant bacteria sometimes arose because of sexual relationships (the sexuality of bacteria having been discovered by the afore-mentioned Lederberg in 1945; until then it had been supposed that bacteria could only multiply themselves through cellular division). So now we also held the key to Griffith's discovery, made 37 years before.

Translocation and deletion

By 1986 the problem had still not been entirely solved, for it was not yet known how the plasmid of the first bacteria 'learned' to produce the necessary enzymes to become resistant to antibiotics. This learning process seems to be similar to that which makes certain insects, such as Californian mosquitoes, resistant to insecticides (see p. 76).

Meanwhile progress in biology, the development of molecular biology and genetic engineering, had opened the way for several discoveries which explained many aspects of genetic recombination.

As a general rule the genetic make-up of an organism can vary without the intervention of external factors, whether by displacement of genes (translocation), or by suppression of genes (deletion). Some of these modifications do not show up as a change in the fundamental type of the organism, others lead to mutations; and finally, some recombinations lead to the death of the organism (lethal recombinations).

These variations are much more frequent at the lower end of the scale of living beings, for example among bacteria and insects, than they are at the top of the scale, among the higher vertebrates. Among the lower organisms a certain number of recombinations occur, not directly in the genes, but on intervention of the plasmid.

These modifications can be multiple and simultaneous, and it has been noted that the frequency of multiple mutations is greater, the closer the recombination sites or loci are together. In addition, numerous external factors — viruses, chemical products, ionizing radiation — can induce genetic recombinations.

Crisis of conscience

Since 1975, the possibility of planned genetic recombination, introduced by the discovery of restriction enzymes (see p. 187), among others, has given rise to several philosophical crises in the scientific world. The Asilomar Conference in 1975 reflected the first real awareness of biologists and political authorities of the issues behind attempts at fundamental modification of living beings, from bacteria to humans. It appears, however, that virtually all attempts at genetic recombination undertaken until now have had entirely therapeutic aims, notably the correction of hereditary diseases and the perfecting of new techniques to combat these conditions.

The mystery of the XX men

One of the most disturbing examples of genetic recombination in humans is that of men with an XX karyotype (that is to say, with two female chromosomes), who ought to have been women and yet are men. It seems that this anomaly may be due to a translocation of fragments from the Y chromosome to the X chromosome. These translocations have a significance far beyond that of an anomaly, since they are produced at sites where, in addition, genetic oncogene sequences (which can produce tumours) have been located. This coincidence would indicate that certain cancers and leukaemias may be due to genetic recombinations.

Germanium

Mendeleyev, 1871; Winkler, 1886

Invented, discovered, rediscovered

The development of semi-conductors was dependent upon the existence of this rare metalloid.

In 1871, in his periodic table of elements, the Russian Dimitri Ivanovitch Mendeleyev claimed that there must be in existence an element with an atomic number of 32 (i.e. 32 protons in its nucleus), which he speculatively called ekasilicon. While analysing argyrodite ore, the German Clemens Winkler discovered, in 1886, mingled in tiny quantities with silver sulphide, a metalloid to which he gave the name germanium, derived from that of his country. This very rare metalloid, only found in the Earth's crust in proportions of from 0.0004 per cent to 0.0007 per cent was industrially exploited after 1945 with the birth of electronics, once its properties as a semi-conductor were discovered.

Germanium was used in the invention and exploitation of the transistor. The first generation of transistors used germanium (although most now use silicon). Germanium is used in military technology: in infrared detectors and as a material from which infrared lenses, periscopes and tank windows are made.

Metalloids
Elements midway between metals and non-metals, they make a diagonal band across the periodic table and are generally considered to include boron, silicon, germanium, arsenic, antimony, tellurium, and polonium.

Present-day uses
- Transistors
- Photoelectric cells
- Current rectifiers
- Lenses with a high index of refraction
- Catalysts of polymerization of polyesters

Dimitri Ivanonich Mendeleyev
Mendeleyev was born in Tobolsk in 1834. From 1866 he was professor of chemistry at St Petersburg. He formulated the periodic law by which he predicted the existence of several elements which were subsequently discovered. Element No. 101 mendelevium is named after him.

Clemens Alexander Winkler
Winkler was born and educated in Freiberg where he later became professor of chemistry. As well as discovering germanium he made important contributions to the study and analysis of gases.

Glycogenic function of the liver

Bernard, 1848

Where energy comes from

By showing that the liver produces glycogen, Claude Bernard made a great step forward in physiology.

Around the middle of the 19th century the existence of glucose was known, a 'sugar' or more exactly a carbohydrate which cannot be decomposed in water. It was known that it circulated in the blood and was used by living tissues to produce heat, a phenomenon whose study had been undertaken by Lavoisier and Laplace. According to their theory it was supposed that the 'combustion' of glucose, about which very little was known, occurred in the lungs.

Then Claude Bernard, founder of modern physiology, carried out a very simple experiment on an animal: he took some blood which was entering the liver and some blood which was coming out of it, analysed the glucose content of each sample and discovered that the first did not contain any, while the second did. Therefore it is the liver which manufactures organic 'sugar'.

More precisely the liver produces a substance which is transformed into sugar and which Claude Bernard named glycogen, meaning 'which produces sugar'. It was to be another seven years before he would isolate this substance; it is a complex sugar stored not only in the liver, but also in the muscles, and which is broken down into glucose by the action of enzymes (see p. 66). Taking his researches further, Claude Bernard went on to measure the amounts of glucose in the blood in different circumstances. He discovered that levels are remarkably constant and that any disturbance is the secondary cause of diabetes (the first cause being insulin deficiency).

The combustion of glucose

Finally, Claude Bernard would demonstrate the role of the sympathetic nervous system, which regulates the size of the blood vessels, therefore the supplies of blood and, with it, of glucose and oxygen to the tissues. It is therefore this system which regulates body temperature, the 'combustion' of glucose occurring not in the lungs but in the tissues.

The significance of these three successive discoveries was immense. In the first place, the glycogenic function of the liver revealed the principle of internal secretions, of which this was the first known example. In the second place, it introduced the biochemistry of muscle and hence of metabolism in general; finally it established the consistency of the internal environment and in addition the very notion of an environment whose disequilibrium leads to illness. Taken together, the three discoveries considerably transformed physiology; but even on its own, the discovery of the glycogenic function of the liver clarified the multiple functions performed by this organ, which until then had remained almost entirely unknown.

> **The founder of physiology**
> Claude Bernard successfully studied areas which were until then poorly understood, such as the role of the pancreas in the digestion of fats, the role of gastric juice, red corpuscles as vehicles for oxygen, the role of the chorda tympani. ... His book *Introduction to Experimental Medicine* really paved the way for modern medicine by rejecting the old doctrines and empiricism in order to replace them with the research of facts which were verifiable and therefore truly scientific.

Gravitation (Laws of)

Newton, 1679–87

The salutary apple

The laws which rule the attraction of one celestial body by another and which are the foundation of modern astronomy were discovered by a single man.

At the end of the 17th century interpretation of the motions of the planets was governed by the ideas of two scholars: those of Kepler on heliocentricity (the Sun at the centre) and those of Galileo on inertia.

The legacy of Kepler and Galileo

In astronomy, understanding of the movements of the heavenly bodies was governed by Kepler's Third Law, which established a link between the orbital periods of the planets and their distance from the Sun, and the mechanics of heaven and earth by Galileo's law of inertia. Despite the innovative character of Kepler's Third Law and the principle of inertia, the academic world of the time was still strongly attached to Aristotelian ideas, and particularly to the concept of the Earth as the centre of the universe. It is against this background that we must consider the originality which led

the Englishman Sir Isaac Newton to discover the principle of universal gravity.

Newton was working in 1669 on the mechanism of attraction and repulsion on Earth. He had approached it first of all from the chemical and physical point of view, studying particles, of the nature of which he could only have had a more or less philosophical or even Aristotelian idea, given the state of science at the time. Suffice it to say that Newton did not tackle gravity from a basis which would be called scientific nowadays.

Hooke and Newton

Sir Isaac Newton had a notoriously touchy and suspicious character as his crucial relations with Hooke, among many other facts, bear witness. Robert Hooke, one of the most eminent members of the Royal Society, had been in conflict with Newton after the publication of the latter's theory of colours (see p. 115), Hooke then being considered the uncontested master of optics and Newton a mere unknown. At the end of 1679 Hooke addressed a letter to Newton in which he informed him of his observations on planetary movement, inspired by the following idea: the planets would propel themselves in space according to a rectilinear movement (ie in a straight line) if it were not for a constant diversion from

---

The famous apple

A popular legend has it that Newton conceived the idea of universal gravitational attraction when hit on the head by an apple. The length of time the scholar spent considering the question and the fairly well-known story of his progress hardly speak in favour of the legend.

this movement caused by a central point of attraction. Newton did not reply to this but embarked on a theory which he published and according to which a spherical body thrown from the top of a tower would fall slightly to the east of vertical, given that the tangential speed of the top of the tower (due to the Earth's rotation) is greater than that of its base. The theory is false, and Hooke did not fail to point this out, adding that, by reason of the Earth's movement, the trajectory of the body would form a segment of an ellipse. Newton, very unwillingly, had to bow to Hooke's argument, which he in turn corrected by postulating that, gravity being constant, the trajectory would be rectilinear. To this Hooke replied that gravitational force is indeed constant but decreases according to the square of the distance from the attracting body.

Two major publications

There we have an essential point, for Hooke was right, but he had only postulated the decrease in gravitational force on purely intuitive grounds. He had indeed applied, in order to reinforce his claim, the third of Kepler's laws which asserts that the movement of the planets is governed by the radii of their orbits, but his calculations were wrong. Whatever the case, Newton must have approached the problem of gravity on specifically mechanical and mathematical grounds. His quarrels with Hooke had very probably helped him to clarify his ideas. He

Shattered reason

In 1678, after a quarrel with Jesuits who claimed that his experiments on the spectrum of light were mistaken, Newton sank into a deep depression which lasted six years. Today we have good reason to think this was due to continuous inhalation of mercury vapours. In 1693 he fell prey again to a classic psychiatric disorder, accusing his friends Locke and Pepys of terrible misdeeds. It has been established that this second attack was caused by the end of the only deep emotional experience of his life, that of his relationship with the young Swiss scholar, Fatio de Duillier.

first put these down in a small treatise called simply *De motu* ('Of movement'), published in 1684, and in 1687 he published his major work, *Philosophiae Naturalis Principia Mathematica* ('Mathematical Principles of Natural Philosophy'), in which he set out three laws:

1. A body remains at rest or in uniform motion in a straight line as long as no force is exerted upon it
2. The rate of change of speed of a body is proportional to the force exerted upon it
3. For every action there is an equal and opposite reaction.

This treatise of celestial mechanics represents original thinking of the first rank: it provided the possibility of establishing the formula of centripetal force exerted on a body to change its rectilinear motion to circular motion by taking as a basis the mass and velocity of the body. It made Kepler's Third Law much clearer. Furthermore, by inserting this formula into Kepler's Law, Newton established that the centripetal force which keeps the planets in their orbits round the Sun must decrease according to the square of their distances from the Sun, and that this law applied equally to the Earth and Moon. He called this law gravity, after the Latin word *gravitas*, 'weight'. He could deduce from this the orbits of the comets and explain the tides. He even set out the bases of nuclear physics, claiming that each particle of matter in the universe attracts every other particle with a force proportional to the product of their masses and inversely proportional to the square of their distances apart.

The publication of the *Principia* was hailed, as might be expected, by the international scientific community, with the exception of some scholars who were appalled by the concept of a force which would operate at distance and which confined itself to Cartesian mechanics. It also provoked scandalized protests from Hooke, who called it plagiarism. It is certain that the old scholar, who had made a mortal enemy of Newton with his arrogant criticism of his treatise on optics, had played the catalyst to Newton's ideas; but it is also quite certain that he did not have the breadth of imagination nor the ability for analysis and synthesis which had allowed Newton to complete Kepler's laws with authority and lay down the foundations of astronomy and modern physics. Vengeful to the end, Newton did not accept membership of the Royal Society until his old enemy was dead.

The Newtonian laws are still applied in contemporary astronomy, remaining no less important for being modified by Einstein, but they are, to a certain extent, seen in the light of a new survey of time and space. This has proved to be valuable in explaining certain anomalies in the movement of planets like Mercury. In 1840 Le Verrier pointed out that the advance of the perihelion of this planet (a property of its orbit) is greater than that which the Newtonian tables forecast and therefore constituted an exception to Newton's law. General relativity, introducing the complex notion of the mass of a body creating a 'curvature in space-time', has exactly accounted for the anomaly discerned by the observation.

The post-Newtonian era

Whatever the circumstances, the appearance of a post-Newtonian celestial mechanics, founded essentially on the observation of phenomena unknown in the astronomy of Newton's time, like expansion of the universe, black holes and quasars, does not invalidate Newton's laws, but extends them in the field of astrophysics.

Hall effect

Hall, 1880

An electromagnetic phenomenon

The carriers of charge in a conductor are deflected by an externally applied magnetic field.

Named after Edwin Herbert Hall its discoverer, the Hall effect is an electromagnetic phenomenon which finds application in electronic devices and which occurs in plasma physics processes such as MHD and astrophysical plasmas. The Hall effect may be observed when a conductor carrying electrical current is subjected to a magnetic field. Charges carrying the current experience a force due to the magnetic field. The direction of the force is mutually perpendicular to the current and the magnetic field, it produces a separation of charge across the conductor which appears as an electric field or voltage across the conductor.

E. H. Hall reported in 1880 his investigation of an attempt to verify or refute James Clerk Maxwell's contention that the mechanical force acting on a current carrying conductor in a magnetic field acts on the conductor rather than the current. Hall argued that if the force acted on the current then an electric field would appear perpendicular to the current and the magnetic field and that the resistance should increase. He observed the electric field using the strip of gold leaf as a conductor, later he also experimented with silver, iron, nickel and platinum. This transverse potential difference (or voltage) is known as the *Hall e.m.f.* (electromotive force).

A useful relationship

The Hall e.m.f. is found to be directly proportional to both the transverse magnetic flux density and the current through the specimen.

$$\text{Hall e.m.f.} = RBJT$$

T is the thickness of the material, J is the current density, and B is the magnetic flux density. R is termed the Hall coefficient.

Applications of the Hall effect

The size and polarity of the Hall coefficient can be used to tell us something about the material of the conductor. For good conductors of electricity like gold, silver, and copper the coefficient is negative showing that the charge carriers are negative. However for some metals such as iron and cadmium it is positive, the charge carriers are positive or 'hole' like because of the complicated band structure of these metals.

Semiconductors show large Hall coefficients which can be positive or negative. The Hall effect is used to diagnose the composition of semiconductor materials

Some uses of the Hall effect arrive from the multiplying property of the relationship which are:

  measurements of electric current
  measurements of magnetic field
  measurements of electric power.

Hall effect devices are used as non contact position sensors and speed sensors. An example of the speed sensor is to detect brake lock up in anti skid braking on automobiles.

As well as metals and semiconductors, electrical conductors include ionized gas plasmas. The Hall effect in a plasma passing through a magnetic field is the foundation of a Magnetohydrodynamic (MHD) electrical generator.

Heavy water

Urey, 1931

The first moderator of neutrons

Rare in the natural state, heavy water played a strategic role in the production of the first atomic bomb.

Studying the spectrum of hydrogen, with the aid of a spectrograph of his own invention in 1919, the British physicist Francis William Aston discovered an apparent error: the spectrum that he had obtained corresponded to an atom with an atomic weight of 1.00756 while the correct weight would have been 1.00777. This difference intrigued several researchers, among them the American Harold C. Urey, who supposed that the presence of a foreign substance in the hydrogen could explain it. So Urey distilled some hydrogen and obtained tiny quantities of an isotope of that element, deuterium or ^2H. The technique of distillation consisted of submitting hydrogen vapours to a pressure sufficient to separate the two elements. In 1931 Urey and his team announced that they had discovered an unknown form of water, heavy water, in fact deuterium oxide or D_2O, the atoms of deuterium combining themselves with those of oxygen in the same way as hydrogen atoms. So the difference in the spectrum of hydrogen was explained.

At first heavy water seemed to be a mere physical curiosity with a molecular weight of 20 (2 for each atom of deuterium and 16 for the oxygen) instead of 18 for ordinary water (1 for each atom of hydrogen and the same 16 for oxygen). Heavy water exists in infinitesimal amounts in ordinary water in the ratio of one part to 6760 of ordinary water. It is formed by the effect of cosmic rays. It is toxic to plants and has curious effects on inferior forms of living creatures whose biological rhythms it lengthens (as was discovered much later).

Nuclear research

In the 1930s heavy water began to command interest for reasons of nuclear physics; it exhibited the property of being able to absorb, to a certain degree, neutrons, which made it a moderator of the reactions of nuclear fission. However, the technique of continuous electrolysis of water required considerable quantities of energy and, in 1940, it was still in its experimental stage. Only Norway had amassed an appreciable stock of it, some 200 litres. With the impending prospect of producing an atomic weapon, the Norwegian reserves took on a strategic importance of the first rank. A government minister, Dautry, decided to send them to France before Germany seized them, and Frédéric Joliot-Curie dispatched them to Britain by submarine in the care of two researchers. This initiative, in association with the destruction of the Norwegian factory, assured the Allies of the exclusive ownership of the world's existing heavy water reserves as well as a considerable advancement in nuclear research.

A lucky error

Paradoxically and providentially the German delay in this area was due to an error of calculation on the part of the atomic scientist Heisenberg, who advocated the use of heavy water as a moderator, even though carbon could easily have replaced it in the research programme that the German physicists had embarked upon in 1940.

Heparin

McLean, 1916; Howell and Holt, 1918

The first anticoagulant

This natural anticoagulant made effective treatment of embolisms (thrombosis) possible.

At the beginning of the century, venous or coronary embolism, that is to say the formation of a blood clot which blocked a blood vessel, remained an accident to which there was no therapeutic answer. In this regard medicine had hardly developed since the time when Fagon, doctor to Louis XIV, advised the use of an apothecary's medications and the application of leeches to 'fluidify' the blood. The use of leeches was not, however, without some medical basis for while sucking the blood these creatures also inject a substance, hirudin, which is slightly venomous but also anticoagulant.

In 1916 the American J. McLean, studying liver cells, stated that they secreted a substance which he had not identified but which, he had discovered, acted against coagulation. Two years later William Henry Howell and Luther Emmett Holt took up and developed his work; noting that blood coming from the liver was particularly rich in this substance, they called it heparin. Subsequent researches would demonstrate that if the liver is really rich in heparin then this substance, a complex sugar, is in fact secreted by all the basophil cells, notably those found in the intestine and the lungs.

The introduction of heparin into treatment, as an extract of the lungs or liver of cattle, was delicate and therefore slow; in fact, in post-operative treatments too much heparin can slow down the healing process, and indeed encourage haemorrhages. It was only in the 1940s that it was isolated in a pure form and applied regularly for circulatory complaints, for example. It is used in the form of calcium heparinate, which is only administered by subcutaneous injection or preferably intravenously, for on the one hand heparin is destroyed by the digestive tract and on the other it is quickly eliminated by the urinary system. It is better therefore to administer it by perfusion and this is done under constant supervision and after testing.

Multiple effects

Heparin works against the transformation of prothrombin into thrombin (see p. 43), and also prevents the action of thrombin on fibrinogen and counters the action of the platelets. It is a new treatment, not only for myocardial infarction (heart attack), pulmonary infarction, phlebitis and embolisms in general but also for numerous inflammatory conditions, ascites (fluid in the peritoneal cavity), pleurisy and exudative rheumatism. Its action on the fats in blood is as powerful as its anticoagulant properties, and for this reason it is also used in cases of excess cholesterol.

Dicoumarin

Heparin is one of the six anticoagulant factors of the blood. Since 1941 the discovery of the antivitamin K or dicoumarin has greatly improved the treatment of emboligenic or clot-forming diseases which began with heparin.

Heredity (Laws of)

Mendel, 1865

The laws of reproduction

For nearly half a century, the laws which governed the transmission of characteristics in living creatures were totally neglected.

One of the most disastrous cases of scientific blindness was the reception given to the laws of heredity.

In 1843 a young man born in Heizendorf, then in Austrian Silesia, who had just two years' experience at the Institute of Philosophy in Olmütz, entered the monastery at Brünn (now Brno). Ordained as a priest in 1847, Gregor Johann Mendel was at first a self-taught scientist, until the abbot in charge of the Augustinian monastery sent him to study at the University of Vienna, where Mendel learned physics, chemistry, mathematics, zoology and botany. On his return to Brünn in 1854 Mendel taught natural sciences at the monastery's technical institute, although he had failed his degree. In 1868 he was named abbot of Brünn.

The pea plants

In 1856 Mendel undertook botanical studies in the monastery garden, pursuing an interest in this science which had begun in childhood on his father's farm. The monastery library was also well endowed with works on horticulture, botany and agriculture, and he added to the collection.

Mendel's experiments focused on pea plants in which he had carefully noted certain constant characteristics such as size, colour or lack of colour in the flowers and leaves, position of flowers on the stem, colour and shape of the seeds, and so on.

Mendel first of all noted that when, for example, he crossed a dwarf plant with a tall one he did not produce a medium-sized plant but a plant bearing the characteristic of one of its parent plants, which he was the first to call the dominant characteristic. However, when he reproduced this plant by self-fertilization the characteristics of the grandparents always reappeared in the same proportion: three-quarters dominant to one-quarter recessive. Furthermore, these second-generation hybrids always comprised one-quarter of individual plants resembling the original pure variety of one of the grandparents, another quarter resembling the pure variety of the other grandparent and the remaining half resembling the hybrids of the first generation. He therefore concluded from this that 'it is now clear that hybrids produce seeds which possess one or the other of two differential characteristics.' He then deduced, with what we can only call genius, that there existed units of heredity, which would later be discovered as genes (see p. 77), which go in pairs and divide in the descendants. He called them

> **Debt to Mendel**
> On the whole, if modern genetics has considerably enriched our knowledge of the transmission and appearance of characteristics, it still owes a considerable debt to the monk who laid down its basic principles.

AA and aa for the original constant (pure) varieties and Aa for the hybrids. The first of Mendel's laws, the principle of segregation, derives from this and states that each parent gives half its units of heredity or genes, which are expressed according to whether or not there is a dominant gene. However, the presence of dominant characteristics is not a constant and in this case the genes readily combine to form hybrids exhibiting features of each of the parents, the combination being made by chance. Using peas which exhibited seven differential features, Mendel calculated statistically the frequency of recurring characteristics in his cross-breeds; he then gauged this frequency in later cross-breeding using his calculations as a basis.

The results of his experiments were read in the course of two lectures which he gave in 1865 to the Society of Natural Sciences of Brünn, which had been founded in 1862 by several of his friends; the text of the lectures was published the following year in the society's bulletin. This bulletin was sent to various scholarly societies throughout the world; Mendel's writing did not arouse the least interest and he died in 1884, completely unknown.

The phenotype

The notion of the phenotype, introduced by Johannssen, would inspire a great many observations and experiments whose repercussions have not yet all been evaluated. Among these mention must be made of the acquired resistance of numerous bacteria to penicillin, a characteristic transmissible by heredity; this poses the question of Lamarckism the idea that acquired characteristics eg fitness as they have passed on to future generations. Neither should we forget the equally acquired and hereditary modification of stocks of penicillin after exposure to rays of light, experiments carried out by Gustaffson in 1937. In both these cases, amongst others, the modification of the genotype enters the area of mutations (see p. 127) whose exploration is still incomplete.

Emerging from oblivion

In 1900, quite independently, three botanists, De Vries, Correns and Tschermak arrived at the same results and conclusions as Mendel; they enquired after possible predecessors and discovered the monk's unrecognized works. A vast series of experiments was to demonstrate that Mendel's conclusions on the mechanism for the transmission of hereditary characteristics applied not only to peas and other plant species, but also to practically all sexed forms of life.

Obviously enquiries were made about other works by Mendel and it was discovered, not without amazement, that on his death his Augustinian colleagues at Brünn had burnt all his papers, judging perhaps that his works on heredity had a whiff of brimstone about them. ... Even more disturbing is the suspicion of relative falsification which would fall upon Gregor Mendel's writings. The calculations which he published are a little too good to be totally true; they do not include the least margin for error, as if they had been given a push in the right direction by sheer genius. If this was indeed the case it only serves to underline the power of his intuition!

Mendel's discovery would be completed and modified by many other developments such as the discovery of chromosome linkage, a concept first presented as a hypothesis by the American T.H. Morgan in 1909 Morgan, who discovered chromosomes (see p. 39) and who had initially doubted the merits of Mendel's theories, postulated in 1909 that two genes placed on the same chromosome could first come together (linkage) then divide in such a way that one gene would find itself on one chromosome and the other gene on another chromosome; this is known as translocation, or crossing over. Defined and verified by many trials this discovery somewhat redeemed the apparently mathematical character of Mendel's laws in that it revealed the possibility of the appearance of characteristics not present in either of the two parents. However, this phenomenon, which is relatively frequent in certain lower animals and in certain plants, is linked in most cases in the human being to congenital defects.

The experiments of Johannssen

Furthermore, experiments completed at the beginning of the century by Johannssen indicate that the characteristics of an individual are not governed solely by genetic distribution, as set down in Mendel's laws.

At the beginning of the century in Britain at the end of a quarrel in which biological statisticians (who held that Mendel's laws did not account for quantitative variations in characteristics) opposed supporters of Mendel, Johannssen distinguished Mendel's genotype (genetic information) from the phenotype (expression of that information) of the individual; the latter is due to the interaction of hereditary characteristics and the environment, and can therefore be variable. In addition, Mendel's laws do not take account of two modifications introduced into the phenotypes by hybridization; a hybrid is not only the passive combination of two genotypes, it is often a lot more vigorous than either of the parents (perhaps because of the interaction of dominant genes), and it is also more stable in different environments.

Lamarckism

A professor of zoology in Paris from 1873, Jean Lamarck is famed for his work on classification and variation. He proposed also that changes in species development are influenced by the environment; his most quoted example is the giraffe's neck, which he thought was the result, over generations, of the animal reaching up for food. He thought that such a change could be inherited, so his ideas were largely abandoned after the work of Darwin and Mendel.

Donkey + *mare* =
mule Donkey (above),
mule (extreme right) and
mare (right). The mule is a
hybrid produced from the
donkey and the mare. It has
a larger head and longer ears
than the horse, and is sterile.

HLA (System)

Dausset, 1950–77

The guarantee for transplants

This complex system by which the organism defends itself has allowed transplant techniques to be perfected and has immensely enriched immunology.

Antibodies, substances which protect the organism against invasion by foreign substances, were discovered early in the 20th century, aided by the discovery of how allergies operate (see p. 14), and by the discovery of blood groups (see p. 31) and agglutinins. Biologists and those engaged in the new science of immunology sensed, however, that there was still a lot to be learnt about how an organism defends itself. From the 1950s onwards numerous biologists pursued the possibility of transplants, study of which was to get underway some 10 years later, but they knew from experiments on animals that rejection of the transplant was practically unavoidable.

Brittany and HLA groups

For some considerable time the unusual frequency of cases of haemochromatosis in Brittany (north-west France) has been noted: this is an anomaly, often running in families, which leads to a harmful accummulation of iron in the tissues. Haemochromatosis reveals itself around the age of 40–60 by a pronounced brown pigmentation of the skin, diabetes, sometimes an enlarged liver prone to cirrhosis, and heart disorders. It has been established that this disease is linked to the antigen HLA-A3.

New antigens

Also in the 1950s, Jean Dausset, future professor at the Collège de France, was specializing in haematology and studying the white corpuscle antibodies present in the blood of women who have already had several pregnancies. He noted that these antibodies, comparable to agglutinins, were only agglutinating the white cells of a certain number of women, carriers of particular antigens (proteins against which an antibody forms); thus he identified a specific leukocyte (white blood cell) group, called the Mac group. He then established the existence of other groups of the same kind and eventually arrived at the conclusion that there existed on the white cells antigens comparable to those on red cells which had been known for decades.

While the antigen–antibody system of the red cells was relatively simple, since outside the four opposing pairs of the A, AB, B and O groups there was only the rhesus pair (see p. 32), the white cell system was shown to be astonishingly complex.

Finally, Jean Dausset established that there were about 30 antigens and therefore just as many antibodies, from which the luck of the genetic draw chose six for each individual. It followed from this that the number of possible combinations in the HLA system was so vast that the chances of finding two individuals with an identical group were virtually nil.

This antigenic system was called HLA, standing for human leuckocyte locus A. Its equivalent has been found in other animal species, for example the mouse, whose parallel system is called H2. This last discovery was due to the Americans Gorer and Snell, who shared a Nobel prize with Dausset in 1980. Finally, the American Baruj Benacerraf discovered that the antibodies of the HLA system were controlled by genes which he had located in a very complex chromosomal area. He was the fourth recipient of the 1980 Nobel prize for medicine.

Improved compatibility

The first practical consequence of these discoveries was as follows: the HLA system is responsible for transplant rejection, so in order to reduce the risk of rejection as much as possible, Dausset and other researchers developed a technique which consisted of trying to find grafting tissue in which the HLA characteristics were as close as possible to those of the recipient. This is how it has been possible to achieve transplants, notably kidneys transplants, which last longer than they did previously.

There was a second consequence too. Each antibody possesses its individual characteristic or idiotype, and it has been shown that an organism can develop antibodies against its own antibodies: these are anti-idiotypes. It is even possible for these

---

Therapeutic benefit

One benefit of genetic engineering, which it is hoped will spread into the therapeutic area a few decades from now, will be in the correcting of anomalies in tissue groups.

---

anti-idiotypes to trigger the formation of yet more antibodies or anti-anti-idiotypes, which are different from the original idiotypes. So, then, an anti-idiotype may find itself faced with different opponents: the idiotype and the anti-anti-idiotype. It is basically a question of a regulatory mechanism which prevents an immune reaction from depriving or inundating the organism with harmful antibodies; but this regulation can itself be disturbed by factors such as chemical products and ionizing radiation; in such cases, a general failure of the immune system may be witnessed. This leaves the organism exposed to disease or, on the contrary, gives it hyperimmunity, a reaction of auto-immunity characterized by allergy.

Auto-immune diseases

The discovery of the HLA system allowed, during the 1970s and 1980s, for the precise statement of a basic theory: that numerous diseases hitherto incomprehensible are, in fact, simply auto-immune diseases. Hyperimmunity brings about overproduction of anti-idiotypes, which end up attacking certain tissues. It is thought that arthritis or degeneration of the cartilage might be an auto-immune disease. The same goes for certain diseases of connective tissue such as systemic lupus erythematosus, and indeed certain forms of diabetes. It might well be that these diseases could, in the future, be treated by restoring the equilibrium of the immune system. We have already been able to establish that a carrier of the antigen HLA-B27 incurs 130 times more risk of developing ankylosing spondylitis. In fact 90 per cent of sufferers from this disease are carriers of this antigen. Many considerations underline the immense importance of the discovery of the HLA system.

Holography

Gabor, 1947; Leith and Upatnieks, 1963

From the laboratory to the museum

The notion of reconstructing a three-dimensional image was inspired by attempts to further develop electron microscopy.

In 1947 electron microscopy seemed to have reached its limits. In fact it had increased the power of the best optical microscopes a hundredfold (to a magnifying power of 100 000) but it was still incapable of elucidating atomic structures.

In contrast to the optical microscope, whose lenses magnify the image resulting from the reflection of visible light on an object, the electron microscope forms the image produced by a beam of electrons transmitted by the object. The greater definition of such images is due to the fact that the electrons have a much shorter wavelength than, and are therefore diffracted far less than, a light ray.

In order to obtain electronic photos of atomic structures it is necessary to use an electron beam with a wavelength of 2 Ångströms. The practical limit was 12 Ångströms ($1Å = 10^{-10}$ metres).

Denis Gabor, an electrical engineer working at the firm British Thompson-Houston on improving the electron microscope, thought that he might be able to obtain increased resolution by examining in coherent visible light a hologram produced on a photographic plate by the

electron microscope. This led to the invention of holography. Gabor described how to record and reproduce the complete optical signal (amplitude and phase) coming from an object. Experiments which he performed using pinhole sources and filters were inadequate to realise the potential of the technique. It was not until the production of the laser as a source of coherent light that Leith and Upatnieks demonstrated the imaging of three dimensional objects.

Laser holography

In normal photography the intensity of light reflected from and emitted by an object is recorded. In holography the phase of light, that is the degree to which peaks of the light waves are in step, is recorded as well. Thus holograms record all of the information in the original light waves and can be used to reconstruct a lifelike image of the original object. Lasers are strong sources of coherent light, that is all the wavefronts coming from the laser are in step with each other unlike ordinary incandescent or fluorescent lamps which have random phases.

Holographic art

Modern artistic productions may use laser beams or lighting effects; they may also use holography as an artistic technique. The museums of holography in Paris and in New York show the best examples of these. There have also been some successful attempts at producing holographic movies although these have been only a few tens of seconds long. There are great technical difficulties which mean that the 3-D images are sharp in the centre of the picture but loss of phase stability leads to distortion away from the centre.

The advent of lasers

The advent of lasers in 1963 gave a real boost to optical holography and signalled a renewed interest in the work of Gabor. Optical holograms are finding increasing applications in optical testing, in credit card security, in industrial design and in art.

Hormones

First hormones: Vulpian, 1856; Takamine, 1901

The body's messengers

*The discovery in the 19th century
of glandular secretions revolutionized
the whole of physiology.*

'Hormone' is the name given to any substance which is secreted by a gland and performs a specific action in certain tissues or functions. In the 1970s the notion of glands, which had until then been fairly limited, was widened to include organs with other functions as well as a glandular one, such as the heart and brain which secrete substances similar to hormones.

Fruitful experiments

The word 'hormone', derived from the Greek *hormân*, 'to arouse', was not invented until 1905 by the Americans Starling and Bayliss, but at least two hormonal functions had already been discovered by then. The first could be said to be the glycogenic function of the liver, discovered in 1848 by Claude Bernard (see p. 83), although this was in fact an enzyme function. The second was the function of the region of the *adrenal* gland called the *medulla*, defined by Vulpian in 1856. At that time experimental surgery on animals was the method used for furthering knowledge of the function of glands and other organs. This was the process used by the Frenchman Brown-Séquard with regard to the adrenal glands in that same year; by the German Schiff for the thyroid in 1859; then in 1883 by the Frenchman Gley, working on the parathyroids and several other glands in 1891. These experiments were fruitful in that they improved knowledge of the glands, but they did not reveal the method, nor, more importantly, the agent — the hormone responsible.

The first hormone

The first hormone to be discovered was adrenaline. Taking as his basis the works of the Englishmen Sharpey-Schafer and Oliver, who, in 1894, had experimentally raised the blood pressure of an animal by injecting it with extract of the adrenal glands, the Japanese Takamine purified this extract in 1901 and isolated the active constituent used by his predecessors. He called it epinephrine, the original name for adrenaline.

In 1914, the American Kendall, who would later discover cortisone (see p. 46), isolated thyroxine, extracted from the thyroid gland. Its formula, however, was not to be determined until 1927 by the Englishman Harrington. (It would be successfully synthesized the following year by

---

Growth hormones

Much of the most significant progress in our knowledge of the endocrinology of human development is due to Professor Guillemin. It was also he who, between 1952 and 1982, discovered three hormones essential to balanced growth: somatostatin, an inhibitor, and somatocrinin or GHRF, an activator for the secretion of growth hormone by the anterior lobe of the pituitary antehypophysis, and finally somatocrin or GRF, a hormone produced by the hypothalamus, which controls all known pituitary hormones and is therefore of immense importance.

Barger.) For a long time thyroxine was held to be the only thyroid hormone; then, after 1952 and the discovery of the much more active tri-iodothyronine, found independently by Roche, Gross and Pitt-Rivers, several other more or less active thyroid hormones were found. This was the conclusion of studies of the thyroid undertaken in 1792 by the Frenchman Fodéré, who was then working on goitre and the disfunctions of the thyroid responsible for 'cretinism' (a term then in wide use); it was taken up again in Paris in 1882 by the Swiss Reverdin, who studied myxoedema or Reverdin's disease, caused by thyroid deficiency.

It seems to be the case that discoveries in endocrinology (the medical science covering the endocrine glands and their hormones) are rarely made starting from square one; they are preceded by a whole body of research material which just fails to be as conclusive as that of the acknowledged discoverer. This fact is illustrated by the discovery of insulin (see p. 107), achieved in 1921 by Banting and Best under the aegis of McLean. In 1925 Collip, who had been the rather unrecognized collaborator of Banting and Best, discovered parathormone, secreted by the parathyroids, which is essential in the metabolism of phosphorus and calcium.

Hormones and pheromones

In the animal world, substances exist which are most often sexual and which act as chemical messengers; their structure, related to that of hormones, has led to them being called pheromones. Their function is to influence the behaviour of other animals of the same species. Since 1980 several researchers claim to have identified analogous substances in humans; they appear to have been found in masculine sweat, and may have the property of regularizing ovarian cycles. This discovery, first proclaimed by the Englishman George Dodd, is disputed by virtue of the fact that human make-up is much too complex to admit of the existence of automatic regulators of behaviour similar to those triggered by animal pheromones.

'Rejuvenating' hormones

Since the beginning of the 20th century a lot of research has been carried out on sex hormones. Interest in these supposedly 'rejuvenating' hormones had in fact been greatly stimulated by the experiments of Brown-Séquard, who had, between 1889 and 1891, advocated the injection of extracts of the testicles and even sperm, with results which were spectacular but misleading. (In spite of this the Russian Voronoff was to attempt, years later in Paris, testicle transplantation, a procedure belonging more to the realm of science fiction than medicine.) From 1923 to 1927 the Frenchman Courrier began a series of discoveries in the field of sex hormones, first by discovering folliculin, the oestrogen (female sex hormone) produced by the ovarian follicle (1924) and not by the corpus luteum. The latter organ does indeed produce a hormone, progesterone, which would be discovered by the Americans Corner and Allen in 1929 and crystallized by the German Butenandt in 1934. Since then the rate of discoveries has speeded up and the complexity of the endocrine system has begun to emerge. In 1931 the same Butenandt discovered androsterone, produced by the testicles; four years later, the German Laqueur, who in 1927 had discovered traces of female hormones in masculine urine, in turn discovered testosterone, also produced by the testicles. In 1929, the year in which Corner and Allen had discovered progesterone, the American Doisy discovered oestradiol, present in the ovaries of pregnant women (and actually capable of being partially synthesized from folliculin). Doisy found this hormone by analysing the urine of pregnant women. As more and more work was done on such urine, the Germans Zondek and Aschheim found yet another sex hormone in it, namely gonadostimulin B, which was to be of particular interest: in 1928 these researchers found that by administering a woman's urine to a female rabbit one could determine if the woman was pregnant according to whether or not the injection triggered the production of eggs in the animal's ovaries.

Knowledge of hormonal function has improved since the discovery in 1920 of correlations between the pituitary gland or hypophysis of the brain and the secretion of sex hormones. It was discovered, however, that the pituitary not only governs sex hormones; it also controls function of the thyroid. Contrary to what had at first been thought, there can be thyroid deficiency, although the thyroid is intact, simply because of the lack of stimulation by the pituitary, that is to say a lack of thyreostimuline, a substance discovered by the Americans Lock and Arar in 1930.

The influence of the pituitary, which was soon to be named 'conductor' of the endocrine 'orchestra' (a title also given to the hypothalamus, part of the brain which controls the pituitary gland), was revealed to be greater and greater as research went on. As in many other fields, long-gone precursors had forged the way. In 1871 the Italian Lovani and in 1886 the Frenchman Marie had noted a link between accidental destruction of the pituitary and growth disorders such as dwarfism, infantilism, and acromegaly usually caused by a pituitary tumour. The mechanism behind this could not be explained at that time as the notion of hormonal secretion had not yet taken shape. Then in 1908 the Romanian Paulesco, the hapless discoverer of insulin (see p. 107), demonstrated that by suppressing the pituitary in an animal, its growth was inhibited, an experiment which was to be taken on board by other researchers several years later (Cushing, Camus and Roussy in 1920). It was not until 1921 that the American Herbert McLean Evans, one of the masters of endocrinology in the first half of the century, discovered that the anterior lobe of the pituitary effectively secreted a growth hormone. He did not isolate it until 1944, with his collaborator and pupil Li, the same person with whom he had first isolated and purified, in 1942, adrenocorticotrophic hormone or ACTH, which stimulates the adrenal cortex. There were, however, dozens of hormones produced by the anterior pituitary still to be discovered.

It was only in the 1950s that a clearer idea of pituitary interactions began to emerge, with a string of ever more numerous discoveries. ACTH was found to function in the manner of a sexual stimulant, and similarly the adrenal cortex was found to act as a sexual gland since it secretes androgens, male sex hormones. Excess androgen can cause virilism in the adult female and an early false puberty in young boys. On the other hand, cortisone, which is also produced by the adrenal cortex, checks the pituitary and re-establishes equilibrium. It is as if alongside the recognized sex gland system another one is in operation.

The role of the emotions

It was also to be discovered in the 1950s that ACTH itself is controlled by other hormones, emanating from the hypothalamus in the brain. This was a major discovery, achieved by the Frenchman Guillemin, discoverer of cerebral hormones and endorphins (see pp. 36 and 65): the anterior lobe of the pituitary only secretes ACTH when there is present an extract of the hypothalamus or of the posterior lobe of the pituitary gland. This considerably altered the rather too mechanistic view of endocrinology: the endocrine system, or at least part of it, is dependent on the emotions, since the hypothalamus reacts to cerebral signals. It is nearly half a century since the American Dale demonstrated that the posterior lobe of the pituitary secretes a hormone which raises blood pressure a (vasopressin) and another (oxytocin) which, amongst other functions, stimulates the contraction of the uterus and initiates labour.

The antimüllerian hormone

It would take an article several hundred pages long to give an account of the complexity of discoveries in the field of endocrinology. In fact an original discovery can be given whole new significance by subsequent finds, as was the case when the parasexual properties of ACTH became known. A particularly telling example is that of the antimüllerian hormone discovered in 1947 by the Frenchman Alfred Jost. This hormone, which only appears on the 37th day of gestation, plays a fundamental role in the determination of sex since it prevents, in boys, the development of the Müllerian ducts which would otherwise lead to the formation of a female genital system. In 1986 it was revealed that this hormone could be used in the treatment of female genital cancers, cancers formed from cells rather similar to those of the female embryo.

The heart's hormone

Over the decades the notion of the endocrine system has been extended. Whilst in the first half of the century orthodox physiology maintained that endocrine secretions were the work of specifically glandular tissues, it has become apparent in the following decades that organs of a non-glandular nature could also secrete substances of a hormonal type. The first demonstration of these cerebral hormones (see p. 36) opened a whole new chapter in endocrinology.

During the 1980s another example was provided by the highlighting of the regulation of the amount of salt in the body through the agency of a hormone secreted by the heart, a muscle which no-one had ever dreamt could have an endocrine function. The possibility can no longer be ruled

---

Anabolic steroids

These drugs have structures similar to the male sex hormones (androgens) but with reduced androgenic activity and increased anabolic activity to increase weight and muscle development. Used clinically to accelerate recovery from protein deficiency, in muscle-wasting disorders and breast cancer, they are also used illegally to promote the performance of athletes and racing animals. They were first used by weightlifters and bodybuilders but have also been used by competitors in all areas of sport. The international Olympic committee requires medal winners to have urine samples to demonstrate that they are drug free. At the 1988 Seoul Olympics the Jamaican-born Canadian runner Ben Johnson was deprived of his gold medal for the 100 metres when it was discovered he had used illegal substances in preparation for the race.

---

out that other tissues exercise an endocrine function by means of cells already known or yet to be discovered. The heart is, without doubt, one of the organs which have been subjected to the greatest number of microscopic examinations by generations of researchers and doctors; and yet it was not until 1951 that the American de Bold discovered in the lining of the auricle (heart chamber) granules which resembled traces of endocrine secretion. It was another quarter of a century before this same scholar stated that the injection of extracts of the auricle lining lowered blood pressure in rats and brought about a greater elimination of sodium in the urine.

The hormone in question was natriuretic auricular factor (NAF). Since its discovery work has been going on to find medical applications for it.

Embryonic hormones

It has been known since the work of the Frenchman Alfred Jost and the Swiss Emile Witschi in the 1930s that the embryo does not secrete the same hormones as the adult. It was thought that many of these hormones, such as the antimüllerian hormones discovered in 1947, disappeared at birth or soon after. The study of these hormones constituted a wide field of research, for several of them, like the female cortexin and the male medullarin, closely conditioned the translation of the initial genetic message, or, in more precise terms, the conversion of the genotype (genetic information) into the phenotype (the characteristics of the organism). It was these hormones which, it was supposed, would be responsible, if disturbed, for sexual anomalies such as hermaphrodites and pseudo-hermaphrodites.

Classification

The abundance of endocrinological discoveries has modified the traditional classifications of the glands, but not the classification of the chemical structure of hormones. Three groups are currently recognized: those which derive from the sterolic nucleus, like the male and female sex hormones and the corticoadrenal hormones; those which are protein molecules or polypeptides, like insulin, glucagon, pituitary and parathyroid hormones; and finally those which derive from amino acids, like adrenaline or thyroxine.

Since the 1920s numerous hormones have been synthesized and since the 1980s advances in genetics have made it possible to produce them in a pure form, as is the case with insulin. The synthesizing of hormones has also meant that it has been possible to modify several of them for therapeutic purposes, for example oestrogens such as hexoestrol, stilboestrol and others. Synthesis has also facilitated the production of artificial oestrogens with multiple properties, such as pregneninolone which is oestrogen, androgen and progestogen all at once. It was synthetic oestrogens which allowed the development of contraceptive pills with very precisely calculated properties.

Two recent discoveries

In 1985 Professor Roger Guillemin, winner of the Nobel prize for medicine and discoverer of cerebral hormones, discovered a new hormone, inhibin, so called because it checks the activity of one of the two sex hormones or gonadotrophins FSH, follicle-stimulating hormone, which acts on the ovary and male genital tract the other being LH, luteinizing hormone, which stimulates ovulation and androgen secretion. Inhibin seemed at first to be an exclusively male hormone, then it was discovered to be present in women too. (In fact there are two types of inhibin, A and B.) In 1986 the same scientist discovered the opposite hormone, activin, which paradoxically, is derived from inhibin. These two new hormones seem destined to play a fundamental role in the treatment of genital cancers, including cancer of the prostate gland, as well as in the treatment of early puberty and the development of a masculine contraceptive.

Induced mutations

Müller, 1927

A half-controlled phenomenon

The possibility of accidental or induced mutation of the genetic capital poses more questions than it answers.

At the beginning of the 20th century, the fascination engendered by the quite recent discovery of genetic laws (see p. 90) led biologists to undertake all kinds of experiments. What particularly intrigued them was spontaneous mutation, then seen essentially from the Darwinian viewpoint. Spontaneous mutations did indeed exist (see p. 127), but they were rare and difficult to observe. A certain number of scholars had reason to think, on the basis of medical experience — still very slight, that X-rays (a form of ionizing radiation) caused profound damage to tissues; so why not to genes? In 1927, the American Müller proceeded to test this in fruitflies. He obtained, beyond all expectations, rates of mutation which could be up to 150 times higher than in the control lines. The experiments were repeated because all the mutations were not fatal, much to the general astonishment; certain were viable and transmissible, and were similar in this regard to spontaneous mutations. The American Robertson had shown, one year before Müller, moreover (Müller's merit was in being the first to achieve scientific mutation), that the phenomenon which prevailed in the mutations was that of chromosomes fusing together. In addition, duplications were recorded, along with ruptures of homologous chromosomes (identical chromosomes, one from each parent, which pair during reproduction) leading to mutations such as translocation, inversion and deletion.

Three triggers for mutations

As the experiments and discoveries went ahead it was established that mutations could be produced in response to three principle factors and one secondary factor: ionizing radiation, chemical actions, and the introduction of a virus into the DNA of cells, along with a change in temperature of the ovum or egg, which only seems to be of importance in the lower vertebrates.

From the medical point of view it has been established that the three primary factors are carcinogens (cancer-causing agents) for the individual adult and both teratogens (agents causing congenital malformations) and carcinogens for the foetus. As far as their influences on hereditary characteristics go, uncertainty reigns, although it is certain that a certain hereditary type of cancer does exist. The influence of viruses in particular remains open to speculation. The hereditary character of induced mutations without any doubt constitutes one of the great mysteries of biology and genetics.

Two strands of thought

Certain geneticists today estimate that in the higher vertebrates, including man, great mutations are fatal, by reason of the extreme specialization of the genetic information. An opposing opinion claims, on the basis of established fact (see p. 127), that mutations held to be minor and already hereditary can accummulate, precisely because they are not fatal, and result finally in the modification of the genotype (genetic information) and phenotype (expression of genetic information). The debate will only be settled by a deeper knowledge of human genes.

Infectious bacteria

Leeuwenhoek, 1673; Davoine, 1850; Obermeier, 1873

The power of 'animalcules'

The identification of the micro-organisms responsible for serious diseases has played a major demographic role.

One of the most eloquent examples of missed discoveries is that of bacteria. Observed for the first time under a microscope by the Dutchman Leeuwenhoek (see p. 193), they remained more or less ignored until their rediscovery by Pasteur. Meanwhile doctors continued to attribute diseases to 'miasmas' or to mystical causes, and remained almost impervious to the notion of contagion. The victims of this ignorance, founded on philosophical beliefs, must be counted in their tens of thousands.

Leeuwenhoek was not in fact content with having discovered bacteria, or 'animalcules' as he called them; he also described them and, as the 400 or so letters to the Royal Society in London bear witness, he established a nomenclature for them which was remarkably well done for a pioneering attempt, including the rod form or bacilli, the spherical form or cocci, the spirilla and many others. It cannot be doubted that if they had been carried further, these observations would have brought about the birth of bacteriology much sooner and led to giant strides in medicine. However, when he included microbes in his catalogue of living beings (they would not be called by that name, first given by the Frenchman Sédillot, until 1878), the great Linnaeus classed them in the 'chaos of the infusoria'. Leeuwenhoek seems to have been the first to grasp the infectious effects of bacteria.

Before Pasteur

The anthrax germ, *Bacillus anthracis*, is most often quoted as the first microbe to have been identified; the names usually mentioned in this connection are those of Pasteur and Koch, who made the discovery at almost the same time in 1876. In fact the anthrax bacillus had already been discovered in 1850 by Davaine and it is fitting here to grant him the honour that he has been denied: he was one of the first, if not the first, to substantiate the part played by germs in diseases. *Bacillus anthracis* was then the first microbe identified in the history of bacteriology. The second is accepted as being Koch's bacillus; but in fact it was the spirochaete of relapsing fever discovered by the German, Obermeier, in 1873.

There follow the leprosy bacillus, discovered in 1874 by the German, Hansen; the gas bacillus of gangrene, discovered in 1878 by Pasteur and Joubert; and gonococcus, discovered in 1879 by the German Neisser. From the 1880s the rate of dis-

---

Bacteriology in Europe and America

Until the beginning of the 20th century all the great discoveries in bacteriology were achieved exclusively in Europe. The hesitant beginnings of bacteriology in the United States took place after the 1900s. It must be said, however, that the Americans have largely made up for lost time.

---

coveries speeded up: staphylococcus and streptococcus were discovered in 1880 by Pasteur; the typhus bacillus in the same year by the German Eberth; mucoid bacillus in the same year again by the Frenchmen Bouchard, Capitan and Charrin; diphtheria bacillus in 1882 by the German Klebs and so on.

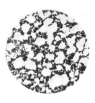

Staphylococcus

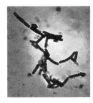

Anthracoid bacterium

Tuberculosis

The notion of infection took a firm hold from that time but it did not so easily conquer established ideas in certain fields, notably that of tuberculosis. The idea that this might be an infectious disease was hardly compatible with the vague notion of 'consumption' which was attached to it. When the German Koch discovered in 1873 the bacillus which bears his name his discovery was rejected for several years; his detractors maintained that this bacillus was not found in consumption lesions, even though the contagious nature of the disease had been established in 1865 by the Frenchman, Villemin. Koch experienced the same sort of difficulty in persuading people to accept that cholera was due to a bacillus: the comma bacillus, which he discovered in 1884 during an epidemic in Egypt.

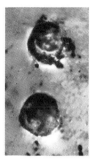

Pneumococcus

Streptococcus

A long line of descendants

In 1885 the German Escherisch discovered the colibacillus and in 1886 his fellow countryman, Weichselbaum, the meningococcus. The Englishman, Bruce, discovered the brucellosis germ in 1887; in 1888 the Frenchmen Chantemesse and Widal discovered the dysentery bacillus; in 1889 Ducrey the soft sore bacillus (rediscovered in 1898 by the Japanese, Shiga); in 1894 the Dutchman Van Ermengem found the botulism bacillus; in 1896 the Frenchmen Achard and Bensaude the paratyphoid bacilla; and in 1899 the Frenchman Thiercelin the enterococcus. Discoveries continued to be made, but at a slower rate. The latest bacterium to be discoverd is *Legionella pneumophila*, isolated in 1978, two years after the serious epidemic of pulmonary infections which had raged at the time of a conference of the American Legion

in a Philadelphia hotel, and which later manifested itself in several other towns.

However, the great number of bacteriological discoveries and the names which are associated with them must not obscure the fact that the masters in this field were Pasteur and Koch, the former for having established unequivocally the role of germs in the triggering of infections, the latter for having established the experimental process which can determine that a given disease is caused by a specific germ. At the present time it seems that the great majority of pathogenic, or harmful, bacteria (which only represent some 3 per cent of all bacteria) have been discovered. Vaccines and antibiotics help to control their pathogenesis relatively well, if we set aside the resistances acquired by genetic mutation (see pp. 79 and 90). It is possible that there may have been major mutations in bacteria in the past: we are still asking what type of germ could have been responsible for the great 'plagues', as the descriptions of these diseases left by the chroniclers bear no resemblance to those accounted for by modern medicine.

Infectious diseases (Role of mosquitoes in)

Finlay, 1880; Ross, 1895; Bignami, Grassi and Bastianelli, 1898

Victories over mosquitoes

*The carrying of malaria and yellow
fever germs by mosquitoes was held
for a long time to be a rather fanciful
theory.*

Malaria is a very ancient disease, and observations about it can be found as far back as the 5th century BC: these are by the Greek doctor Hippocrates, who distinguished five types (of which some were perhaps recurrent fevers). It seems to have raged mainly in the East and Far East, only appearing in the Americas after the arrival of Columbus, since malaria epidemics were recorded in the New World from 1493 onwards.

The *Plasmodium* protozoon

The use of quinine, an extract of cinchona bark, is less ancient; if it was quite widespread in 1700, the causes of the disease themselves remained unknown. The first person to claim publicly that the transmission of germs was due to mosquitoes was the Italian, Lancisi, who also drained large parts of the Pontine swamps, breeding grounds for malaria. It was not until 1880 that the Frenchman, Laveran, isolated the agent of the disease, the *Plasmodium* protozoon. Infection of the blood seemed to be the most likely hypothesis.

In 1892 the British scientist Ross had the idea of studying a mosquito which had just sucked a patient's blood, and he examined the stomach contents: there he discovered the protozoon. In 1895 he proceeded with an original experiment which consisted of putting birds affected with malaria into the

same cage with unaffected birds; contagion did not take place. Ross then introduced mosquitoes and the healthy birds became ill. It could be said, therefore, that Ross was the first to discover and establish the role of mosquitoes as agents of the disease.

The female mosquitoes

What happens in animals does not always happen in man, for there are some diseases which are not intertransmissible. The discovery of the transmission to man was made

---

'Contagionists' and 'miasmatists'
Although the blow dealt by Pasteur to the theories of spontaneous generation of microbes was to prove fatal in the long run, the existence and reproduction of microbes having been duly verified under the microscope, and although diseases were no longer attributed to 'fate' or to obscure 'moral' powers, the quasi-theological debate between 'miasmatists' and 'contagionists' was to last much longer. The contagionists pleaded for the material and verifiable propagation of microbes by dirty hands, insects and infected matter. The miasmatists, on the other hand, invoked 'malign vapours' given off, for example, by swamps; they held back acceptance of Finlay's scientific explanation for several years.

by the Italians Bignami, Grassi and Bastianelli, in 1898. These three researchers took the bold initiative of having a man infected by mosquitoes and were then able to establish the life cycle of the parasite in the blood. They were also the first to establish that the *Plasmodium* is transmitted exclusively by the female of the *Anopheles aegyptii* type.

It was not until 1948 that the Englishmen Garnham and Shortt established that the parasite accomplished its reproductive cycle not only in human blood but also in the body of the mosquito.

Yellow fever

The discovery of the specific role of mosquitoes in the transmission of malaria has rather obscured the earlier discovery of the general role of mosquitoes as vectors (carriers) of disease. It was in fact in 1881 that the Cuban, Finlay, claimed, on the basis of formal clinical proofs, that yellow fever was transmitted by a mosquito, *Aedes aegyptii*. Strangely, no-one paid much attention to his work for nearly 20 years. (Finlay only published his thesis in 1886.) At that time in fact the theory of 'miasmas' had not really been abandoned and it was supposed that yellow fever was transmitted by germs present in the air or in water. It was Ross's discovery which finally brought some credit to Finlay's theory. However, the

---

Mosquitoes against mosquitoes

The spread of *Plasmodium* strains resistant to available treatments (see p. 169) inspired an original solution in 1986, but one yet to be put to the test. Given that there exist species of mosquito which destroy the *Plasmodium* in their digestive tubes as opposed to those which harbour the disease, Frank Collins, an American researcher, envisaged introducing these mosquitoes widely in the natural world in the hope that their genetic characteristics would be communicated to other groups. However, there are 65 species of mosquito which transmit malaria and four different types of *Plasmodium*; we are dealing here with a long hard task.

Americans, who occupied Cuba in 1900 after the Cuban-American War, believed so little in Finlay's thesis that they took the considerable risk of inoculating people with this serious disease. The Americans Reed, Carroll, Lazcar and Agremonte tested the infectious capacity of the mosquito by injecting volunteers from the occupation forces, civil employees and members of the expedition. In 1901, finally, the report of the American Commission confirmed Finlay's work. The agent of yellow fever had been discovered in the blood of the sufferers; it was a virus called amaril, carried and transmitted by mosquitoes. Measures were then taken to combat the insects: drainage of swamps, elimination of stagnant waters, and destruction of larvae by fire.

Other diseases

Once admitted, the role of mosquitoes as vectors of infection allowed for the discovery of the causes of numerous other diseases. These include the lymphatic filariases from *Wuchereria*, dengue fever and viral encephalitis also transmitted by mosquitoes; typhoid, purulent eye infections, tuberculosis, staphylococcal infections, dysentery, cholera and trachoma, transmitted by the house fly; anthrax, transmitted in certain cases by the *Stomoxys* fly; sleeping sickness or trypanosomiasis, transmitted by *Glossina* or tsetse flies; onchocerciasis, also called river blindness, transmitted by blackflies; numerous myiases, intestinal, orbital, cutaneous, subcutaneous, rampant and so on; filariases, transmitted by gadflies; typhus, transmitted by lice; plague, transmitted by fleas. . . . The list of diseases transmitted by insects is certainly a very long one.

Prevention by improved hygiene, recourse (excessive at first) to insecticides, and then from the 1970s onwards biological measures — pharmacology and vaccines based on the discovery of the role of insects in the spread of diseases — have all helped, since 1980, to get these diseases on the run and indeed to eliminate many of them. We have come a long way since the theories of 'miasma'.

Infrared

Herschel, 1800

The hottest light

The discovery of this invisible radiation, situated beyond the red end of the visible spectrum, has brought numerous refinements to astronomy and optics.

Taking up Newton's discovery of the visible light spectrum (see p. 115), the English astronomer Sir William Herschel (a German by birth) had the idea of measuring possible differences of temperature between the bands of the spectrum. He discovered that light in fact gave out heat but, more remarkable than that, he also discovered by accident that, beyond the red band, which marks one of the limits of visible light, the heat given out was much greater. He concluded from this that there was energy coming from the Sun in wavelengths far greater than those which the eye could see, and that this energy, just like visible light, was part of the electromagnetic spectrum.

Beyond red

Herschel also deduced that this radiation was the same radiant heat already studied by Newton, and stated that it could be reflected or refracted just like visible light. The Swiss scientists Saussure and Pictet-Turretin verified and confirmed this assertion in 1803. In 1840 William Herschel's son, John, established that within the infrared band less sensitive zones existed. In 1879 the Frenchman Mouton measured its wavelengths using a thermo-electric couple which had been invented in 1833 by Nobili and then perfected by Melloni. The use of highly sensitive photographic plates helped to verify, during the 1880s, that infrared was to be found in the extension of visible light beyond red.

For several decades hardly any practical developments resulted from Herschel's discovery. The first spin-offs came in the 1920s, when the Americans Coblentz, Pettit and Nicholson discovered that the stars also emit infrared rays and that, the redder they are, the more they emit. It was not until 1948, when radiation with wavelengths as long as two microns (0.002 mm) were detected from dying stars, that infrared astronomy made a serious beginning. It was then already known with the aid of a spectrograph and sensitive films that many molecules radiate naturally at infrared wavelengths.

Infrared detectors have many applications, both military and civil. They are used in night and smoke vision systems (for fire fighting), intruder alarms, weather forecasting and missile guidance systems.

The quantum theory

The discovery of the infrared band has had very important consequences, including the development of the quantum theory, based on the study of thermal radiation. The basis of the quantum theory was the idea of the 'black body' advanced by the German Kirchhoff in 1859. This concerned an imaginary body which completely absorbs all light or heat radiation falling upon it. Another German scientist, Max Planck, was able to interpret experimental observations of thermal radiation by invoking the quantization of energy states within a black body.

Insulin

Banting, Best, Macleod, 1921

The anti-sugar

The discovery and extraction of this hormone have gone far beyond the narrow field of diabetes.

Diabetes is a disease as old as mankind; its causes remained totally unknown until the beginning of the 20th century. Towards the end of the 19th century, however, several doctors began to suspect that this condition had a functional origin. Diabetes means an abnormal excess of urine; it manifests itself in an inability to assimilate carbohydrates. There are in fact several types of diabetes. Symptoms may include chronic skin lesions and damage to the retina, heart and nerves. In 1870 Bouchardat, a doctor, claimed that rationing and forced exercise caused by the siege of Paris led to the disappearance of glycosuria (the presence of excessive quantities of sugar in the urine) in certain patients.

Islets of Langerhans

No connection had yet been established between diabetes and the role of the pancreas. In 1869, the German Langerhans had indeed discovered two distinct types of cells in the pancreas: the acinus glands, which produce digestive enzymes which reach the duodenum via the pancreatic duct; and islets of cells, which the Frenchman Laguesse later called islets of Langerhans whose function were completely unknown.

The first link between diabetes and the pancreas, which must be considered as a discovery in its own right, was established in 1889 by the Germans Minkowski and von Mering at the University of Strasbourg. These two researchers, who were studying the role of the pancreas in metabolism but who had, until then, only carried out their experiments on birds, proceeded to dissect the pancreas of a dog. Their aim was to clear up a point of disagreement between them, which concerned the necessity of pancreatic juices in the digestion of fats. To their surprise the dog suffered from polyuria, that is to say, it produced an abnormally high quantity of urine, one of the signs of diabetes. Analysis revealed that this urine contained 12 per cent sugar, a quite excessive amount, indicating diabetes. Minkowski and von Mering had confirmed that the pancreas plays an essential role in the origin of diabetes.

Naturally, they explored this discovery further. They ligated (tied) the pancreatic duct of another dog to see whether stopping the pancreatic flow to the duodenum did indeed cause diabetes. Surprise number two: the dog suffered from digestive troubles, but not from diabetes. Supposing the ligatures to have been faulty, the Frenchman Hédon repeated the experiment, this time shifting the position of the pancreas completely and transplanting it under the animal's skin. There was now no way that the pancreatic juices could reach the stomach, but still the animal did not suffer from diabetes; that only began with the surgical removal of the organ. It appeared, therefore, from this second experiment, that the pancreas has two functions: its external secretions, discharged into the duodenum, play a part in digestion; and its internal secretions, discharged directly into the blood, control the metabolism of carbohydrates. So it could be said that two quite separate discoveries, one by Minkowski and von Mering and the other by Hédon, paved the way for the discovery of insulin. A third discovery of lesser importance was made in 1901 by the American, Opie, who discovered that there was indeed a relationship of cause and effect between diabetes and the destruction of the mysterious islets of Langerhans.

Elusive insulin

During the next few years progress alternated with discouragement. Attempts were made to treat people in many experimental ways, using extracts of the foetal pancreas and of the pancreas of fish (without results), and using adrenaline (with encouraging results but with serious toxic reactions). A great many researchers tried to obtain pancreas extracts which might contain the mysterious substance hypothetically produced by the no less mysterious islets of Langerhans: the results were either contradictory or disappointing. Insulin, the name given to the substance in 1909 by the Frenchman De Meyer because it was produced by the islets, remained elusive. In 1913 discouragement set in and diabetes continued to be treated by the age-old method which basically consisted of ... opium! In one American hospital a child actually died of hunger while being treated for this disease which had been attributed to too rich a diet.

Banting's inspiration

The medical world had all but abandoned its researches into this subject when, on 30 October 1920, during a sleepless night (legend has it, wrongly, that it happened in a dream), Frederick Grant Banting, a Canadian doctor whose career had not been particularly promising and who held a junior post at the Western University of London, Ontario, read a macabre article. It was an account of an autopsy. In it the author, Barron, described a rare case of pancreatic stones which had blocked the pancreatic duct, while the islets of Langerhans were unaffected. Banting had a

flash of inspiration: if the pancreatic duct of an animal were ligated and a few weeks allowed to pass so that the organ degenerated, then undamaged islets of Langerhans could be obtained, from which it would be possible to extract the internal secretion. Curiously, no-one had ever thought of this before.

Banting submitted his idea to a leading Canadian physiologist, James John Rickard Macleod, who was not exactly over-impressed by Banting's knowledge but who did find the project interesting. However, the experiment had to be carried out in Toronto and so, at his own risk, Banting had to leave his London post. The project did not in fact get under way until the summer of 1921 and its beginnings were inauspicious, largely due to Banting's inexperience. Macleod had selected Charles Best to be Banting's collaborator. In June and July 1921, having successfully obtained atrophied (degenerated) pancreases with undamaged islet of Langerhans, Banting and Best ground them in Ringer's solution (distilled water and salts) and injected the filtrate into animals rendered diabetic by removal of the pancreas. One hour later the sugar levels had decreased by 40 per cent.

These experiments, carried out during Macleod's absence on holiday, had to be repeated time and time again for verification and measurements. Meanwhile Macleod had assigned Banting and Best another collaborator, James Bertram Collip. Collip was an excellent technician, who first discovered insulin shock in animals caused by too-high doses, and who later thought up the method of preparation

---

Clashes of character

Impulsive, inspired and very obviously a genius, Banting failed to gain the sympathy of his future patron Macleod at the time of their first meeting, for Macleod himself was cool, reserved and formal. The fact that, on top of all that, Banting was an orthopaedic doctor and not a physiologist, would not incline Macleod to be indulgent; it is a miracle that he did finally agree to sponsor Banting and allow him to carry out his research under his aegis.

by fractional precipitation of alcoholic extracts of islet of Langerhans, which can safely be injected into humans.

In January 1922 the first trials on a human being were carried out. The first subject was a young boy, Leonard Thompson, who was dying and who recovered. The international response was huge and emotional. In 1923 Banting and Macleod received the Nobel prize for medicine, one of the most disputed awards ever. Much has been written about this award, for Banting, who had conceived of the experiments which led to the discovery of insulin, had carried them out while Macleod was away on holiday; but Banting had chosen Macleod as a patron and it was this patronage which had enabled him to complete his work. The worth of Best's and Collip's contribution also went unrecognized by the Nobel judges, although the former had collaborated very closely and to great effect on the difficult early stages, and the latter had invented the technique for purifying the

---

Generosity in symmetry

Just as Macleod shared his Nobel prize with Collip, whom he considered to have been unjustly neglected by the Nobel jury, Banting shared his with Best, whose talent as a practical operator had contributed greatly to the discovery of insulin and whom the same Nobel jury had also neglected. It is said that that Banting had a terrifying outburst of anger when he learned that it was Macleod who was to share the Nobel prize with him.

---

pancreatic extracts which made possible the first trials on man just a few weeks after the tests on animals. The matter did not rest there, Banting feeling dispossessed of his discovery by Macleod, and Macleod feeling just as unhappy with the Nobel jury's decision. Very generously, Macleod shared his prize with Collip to compensate for the injustice committed in his regard. Finally, a Romanian researcher, Nicolas Paulesco, wrote some very heated letters of protest to the Nobel jury, which had behaved in a rather off-hand manner towards him. In 1921, the same year in which Banting and Best completed their first tests with extracts of the pancreas, Paulesco published convincing results of similar tests with what he called 'pancreine'. It has often been deplored that these trials were not taken into consideration by the Nobel prize jury.

The fortunes of insulin

For years insulin was extracted from animal pancreas, notably from cattle. Since 1964 when Panayotis and Katsoyannis succeeded in synthesizing it at Pittsburgh University, it has been the synthetic form which has been administered. Since 1983 the manufacture of insulin by genetic engineering has been undertaken. The discovery of insulin has enabled the causes of diabetes, which are numerous (there is, for example a renal diabetes called diabetes insipidus), to be distinguished and the essential aspects of the metabolism of carbohydrates and lipids (fats) to be clarified.

Ionosphere

Marconi, 1901; Appleton and Barnett, 1925

The genius of radio

*This strongly ionized layer of the
atmosphere was only discovered
following a bet.*

In 1894, a 20-year-old Italian, Guglielmo Marconi, read Hertz's obituary in an Italian scientific journal (see p. 178). It had been written by the young Marconi's physics teacher, Righi, one of the researchers who had helped to establish the electromagnetic nature of Hertzian waves. Marconi hit upon the idea of using these waves in telecommunications. Equipped with a Branly coherer which he had improved, he erected — and this was of course an invention — two antennae, one a transmitter and one a receiver, to improve the aerial transmission of waves; he then installed a transmitter-receiver circuit in the hills around his home at Pontecchio near Bologna. He established — and this was a vital point — that it was possible to receive Hertzian waves at a distance of three km, that distance being over a hill. It was possible therefore to communicate at a distance.

where else being relayed by wire). In front of Her Majesty's Army and Navy officers he proved that it was possible to transmit messages over three km, then over more than seven km. At the beginning of 1897 he managed to get a message from South Wales, across the Bristol Channel to Brean Down in Somerset, a distance of some 13 km.

He founded his company, with a transmitter post on the Isle of Wight, transmitting essentially meteorological messages to ships entering the English Channel, making for Southampton. Following these modest beginnings he equipped a tug-boat with a 12-metre antenna which allowed Queen Victoria to be in telegraphic communication with her son, the Prince of Wales, who was cruising in the royal yacht off the Isle of Wight. In 1899 he established the first telegraphic link across the Channel between Wimereux and Dover.

A British first

Marconi informed the Ministry of Posts and Telegraphs in Rome of his discovery — for such it was, even if it derived from that of Hertz in its technical angle — but the Ministry was not interested. Marconi then calculated that a maritime power such as Britain might be more receptive to the practical implications of his discovery. He went to Britain in 1896. The British quickly understood the importance of what was in fact radio, or wireless telegraph by morse code (the form of telegraph in use every-

> **Stroke of luck**
>
> Marconi would never have been able to make calculations on the propagation of radio waves across the Atlantic, thanks to a possible electrified atmospheric layer, for these were way beyond the competence of the age. It would appear that the young scholar acted on a hunch with results which were in fact remarkable both for him and for science.

Mass production

Mass producing induction coils and improved coherers, Marconi had, by 1900, equipped a large part of the British, German, Italian and French fleets. There was a growing risk of interference, but Marconi invented a method of control for the frequencies of transmission, which allowed for broadcasts, still in morse, on determined frequencies. At that time transmissions could be made over about 75 km.

Achieving the impossible

It was obvious that Marconi was going to attempt transatlantic communication. At first this seemed to be impossible by virtue of the curve of the Earth's surface. Hertzian waves spread out as they travel, and they would perhaps be lost in the high atmosphere. Armed with an obstinacy which, in hindsight, bordered on absurdity, and having certain ideas he wanted to try, Marconi nevertheless attempted the impossible. He installed an antenna at Cape Cod, Massachusetts (his first was destroyed) and another at Poldhu in Cornwall some 3000 km away. This installation cost him the then considerable sum of 50 000 dollars. It is possible that he was encouraged by the hypothesis put about since 1880 of an electrified atmospheric layer; this alone was capable of explaining the aurora borealis (Northern Lights) and certain magnetic anomalies. So he undertook to transmit from Cape Cod to Poldhu a single morse letter, S. On 12 December 1901, shortly after midday, he recorded the letter S; he

was totally incredulous. He had this verified by his assistant: it was no mistake, the impossible had been accomplished.

The news caused a sensation and quite a backlash. It was questioned with considerable acrimony, as it posed a threat to the interests of those companies owning the dozen or so transatlantic cables then in operation.

A reflecting sphere

Scientifically, the discovery was barely plausible. Poincaré's calculations on diffraction even made it 'impossible.' Experts the world over studied the affair, the Englishman Heaviside, the American Kennelly, and the Japanese Nagaoka suggesting that the Hertzian waves had passed round the Earth by being reflected from higher levels of the atmosphere which would have been ionized. This was, moreover, the opinion of the Frenchman Blondel in 1903 and of Poincaré himself. The hypothesis was confirmed experimentally in 1925, when the Englishmen Appleton and Barnett observed an interference pattern between a wave theoretically reflected off the ionosphere and another directly transmitted to the receiver.

In exploiting a discovery and attempting the impossible, Marconi had discovered the ionosphere, without which there would be no intercontinental transmission of radio waves. Marconi had opened the era of worldwide telecommunications and also, by the indirect means of interference of solar origins, that of radio astronomy.

Isomerism

Boutlerov, 1861

One of the foundations of organic chemistry

The existence of bodies possessing the same elements, but different properties, was a revelation in chemistry.

Since the beginning of the 20th century numerous chemists have come up against a strange problem: certain substances seemed to be composed of the same elements — carbon, hydrogen, sulphur, chlorine, and so on — but they exhibit different properties. The Frenchmen Laurent and Gerhardt, in attempting to explain this, proposed 'typical formulae' in which the proportion of atoms from each substance would be respected.

In this way the formula for water, H_2O, where two atoms of hydrogen are counted for each one of oxygen, could be written

$$O \begin{cases} H \\ H \end{cases}$$

and would allow for different formulae such as:

$$O \begin{cases} H \\ H \end{cases} \quad \text{or again:} \quad O_2 \begin{cases} H_2 \\ H_2 \end{cases}$$
$$O \begin{cases} H \\ H \end{cases}$$

or even:

$$O \begin{cases} H \\ H \end{cases}$$
$$O \begin{cases} H \\ H \end{cases} \quad \text{or} \quad O_3 \begin{cases} H_3 \\ H_3 \end{cases}$$
$$O \begin{cases} H \\ H \end{cases}$$

This theory was to lead to the idea of valencies, which postulated that the elements have a specific 'combining power'. (For example, one atom of carbon can combine with up to four atoms of hydrogen and has valency 4.)

This notion was to be formulated separately by the Englishman Couper and the German Kekule. Paradoxically, Couper, who set it out clearly on an atomic basis for which he was unfortunately unable to provide the proof (ingeniously suggesting that carbon, for example, could combine with itself to form carbonized chains) was much more coolly received than Kekule, who published the same theory far less clearly. Couper eventually went mad.

A master stroke

In Kazan, the Russian Boutlerov took up these theories and introduced the idea of structure, which is the one which prevails today and which authoritatively explains the links. Armed with this theory, published in 1861, he discovered tertiary alcohols in that same year. He demonstrated theoretically that two types of butane (or three types of pentane), for example, could be formed from the same atoms, but would have different properties, properties also discovered by him. In 1866 he synthesized isobutane, and in 1868 came the master stroke; he discovered that unsaturated organic compounds contained multiple carbon bonds. Organic chemistry had just taken a great step forward. The illustrious Berzelius, who had refused to entertain either the idea of structure or that of different properties for substances of the same composition but different structures, bowed to the weight of the evidence and christened the phenomenon 'isomerism'.

Lead tetraethyl

Kettering, Midgley and Boyd, 1921

A performance motorfuel

The addition of this compound to petrol considerably altered car engines.

At the beginning of the 20th century the motor fuel used in cars was a spirit distilled from petroleum. The motor industry, which was trying to find a way of increasing the power of its engines and, in order to do that, raise the rate of compression before ignition from four to six, turned away from this fuel in favour of a spirit obtained from cracking petroleum, a chemical technique which involves subjecting the raw petroleum to high temperature and pressure in the presence of catalysts in order to break it down into its constituent components.

This allowed for the use of much more convenient amounts, which also possessed a distinctly higher octane number. The old, extremely smelly, distilled fuel gradually disappeared from the market.

The introduction of the spirit from the cracking process first of all called for greater precision in the casting cylinders, which was resolved during the years 1918–20. Then certain annoying anomalies appeared in the running of the engines; rattling in the valves, knocking against the cylinder heads and vibration in the gear box, followed by a sudden loss of power in the engine and serious deterioration of several components. This was a mysterious phenomenon and several researchers applied themselves to it, among them the American Kettering, Vice-President and Director of Research at General Motors, assisted by two other technicians, Midgley and Boyd.

The chemistry of mineral oils had not then reached the level of development which would be achieved several years later and, cracking having only just appeared on the scene, the properties of the components from it (such as paraffin and the aromatic forms) were not yet thoroughly known.

A hypothesis

Kettering's team was therefore in the business of breaking new ground. After several months of research it put forward the following hypothesis: the combustion spark causes the formation of an ultraviolet arc which triggers a chain reaction, responsible for the detonation. In this hypothesis, the cyclobutanes formed during this reaction would not burn correctly. The researchers then thought of adding to the spirit a product which would screen the ultraviolet arc and be capable of dissolving with the spirit. Naturally they thought of lead, and after numerous trials they perfected lead tetraethyl spirit, which completely satisfied their requirements. In 1923 the American Ethyl Corporation went into commercial production of leaded petrol.

A glass engine

The researchers' theory intrigued the engineers of the Ford Motor Company who, in 1931, had Corning Glass produce an engine of special glass which would allow them to study easily the different stages of the cycle aspiration of vaporized air, compression, ignition, power stroke, working of the valves, etc. When the engine was completed (it was shown in the American Pavilion at the Universal Exhibition of Arts and Technical Sciences in Paris in 1937) it revealed that there was no ultraviolet produced on ignition.

Knocking

The engineer Henri Weiss wrote that in an engine the advance of the spark to dead centre is adjusted in such a way that combustion is complete at the moment of the most effective power stroke. This ignition is therefore regulated in terms of a speed of flame propagation of some tens of metres per second. If this speed is multiplied several times, as is the case with petrol from cracking, there is an imbalance between the chemical rhythm of combustion and the mechanical rhythm; hence the anomalies in engine function. The normal flames of combustion are replaced by ones called detonation flames, which are released ahead of the flame line, in the part already subject to a strong compression because of the dilation of the part already burnt. The generation of a detonation flame being linked to the composition of the motor fuel, the first concern was to establish relatively anti-detonating mixtures, and then to inhibit the formation of products of addition (born out of the fixing of oxygen on hydrocarbons at relatively low temperatures): in fact, these products of addition favour the production of a detonation flame. So there are two fundamental causes of detonation flames which cause knocking.

Lead tetraethyl has the advantage of being anti-catalytic and that is why it was effective when Kettering's team tested it, but on false bases.

> **Since 1923**
> It was lead tetraethyl that facilitated the great expansion of the car industry worldwide from 1923 onwards, making possible the production of more and more powerful high-performance engines.

The use of lead tetraethyl therefore really represents a lucky discovery or, to put it differently, a perfect solution to a badly presented problem. Nor is it the only one of this kind in the history of discoveries.

Efficiency and pollution

Obviously it would be denying the skills of Kettering, Midgley and Boyd to suppose that they did not reconstruct exactly the elements of the problem after their initial fruitful mistake. That required many long months of research and numerous photographic studies, made easier by the Corning glass engine. It was these studies which allowed them to examine the different workings of the flame and penetrate the chemical problem of combustion. The improvement in car petrol enabled the change to take place from a compression rate of 4.4 in 1925 to one of 9.5 in 1958, with an improvement in output to the litre of 60 per cent. It was the work of the team mentioned above which allowed for the establishing of the optimal levels of addition of lead tetraethyl.

It was from the 1960s onwards that consideration began to be given to the harmful lead emitted into the atmosphere from car exhausts.

> **Lead tetraethyl no longer needed**
> Initially the adding of lead tetraethyl to petrol enabled car engines to run more efficiently. This growth in efficiency led to research which increased it again, notably from the 1950s on. This led to the production of engines which provided very satisfactory efficiency, even without lead. Since the beginning of the 1980s lead tetraethyl has been dispensed with in the production of petrol in many countries and it can be foreseen that by the end of the century this additive will have been totally abandoned, for ecological reasons, since it is a dangerous pollutant.

Light spectrum

Newton, 1666

The cradle of astrophysics

*The separation of white light into
seven colours, the basis of a great
chapter in the history of optics, was
at first a mere curiosity.*

It was in 1666, as he reported in his work *Opticks*, that Sir Isaac Newton suddenly realized that light falling on a glass prism is separated into seven bands of different colours. He repeated the experiment in a rather more thorough manner, by piercing a hole in one of his window shutters to obtain a more limited light beam, and by placing a sheet of paper in the path of the diffracted beam, so that he could note the order of the colours. In 1704 he was to follow up this observation on soap bubbles, which indicated that several thin transparent layers produce the characteristic diffraction of iridescence.

The theory which Newton built from his observations was the fruit of great insight, and his principle has never been questioned: light is a heterogeneous compound of rays of different wavelengths, diffracting at distinct angles and each causing a different visual impression. He was the first to postulate that a rainbow is the product of refraction. Convinced of the impossibility of correcting the chromatic aberrations in lenses, he designed and constructed the first of the telescopes with mirrors.

No-one at the time understood Newton's theory: Hooke, one of the most celebrated members of the Royal Society, did a rather ill-founded criticism of it and the other physicists gave it no better a reception. The English Jesuits even claimed that the experiment on which Newton based his theory had been badly conducted.

Nearly a century and a half elapsed before this curiosity of science was taken up again. It was, in fact, not until around 1800 that the British astronomer William Herschel thought of measuring the thermal intensities of the solar spectrum and found that the release of heat continued beyond the red extremity of the spectrum: he had discovered infrared radiation (see p. 106).

Spectroscopy

Spectroscopy, an essential fruit of Newton's discovery, consists of the study of the spectra of different light sources; it was born thanks to the observations of the German Joseph von Fraunhofer. Systematically analysing the solar spectrum, in 1814, on the information of an English scholar, Wollaston, who had detected dark lines in it, Fraunhofer counted exactly 754 of these dark lines, which are still called Fraunhofer lines. The same scholar was also the first to have

---

Newton, father of colour printing

Almost all colour reproduction in modern printing is done with only three of the seven colours of the spectrum: blue, red and green. In fact, these contain sufficient variety of wavelengths to reconstitute all the colours of the spectrum, thanks to the superimposing of coloured screens. It can therefore be said that Newton is the (distant) father of contemporary printing techniques.

the idea of measuring the specific wavelengths of each band, thanks to a rudimentary diffractometer that he built, the first of its kind.

Fraunhofer's discovery was of major importance, but it was not recognized until much later. The cause of this was principally technical for it was impossible to develop astronomical spectroscopy without large mirror telescopes, and it was impossible to construct these telescopes without large blocks of glass, which no-one yet knew how to manufacture (see p. 22). Many small mirror telescopes existed, but they were not suitable. It was only in 1856, when it became known both how to make large lenses and how to lay a uniform film of silver on the reflecting face, that astronomical spectroscopy, the essential application of Newton's discovery, could at last take off. Six years later, the Swedish physicist, Anders Jonas Ångström, undertook a successful analysis of the chemical composition of the solar spectrum and established the unit of wavelength which now bears his name.

Techniques developed more and more rapidly and, from 1887 to 1896, the American Henry A. Rowland published a solar spectrum of very high resolution as well as a catalogue of the wavelengths and of the intensities of each band of this spectrum. It was, at the end of the day, a magnificent development of Newton's lucky discovery.

Meanwhile another astronomer, the Englishman William Huggins, was compiling an inventory of the spectra of differ-ent luminous celestial bodies — stars, nebulae, comets, planets, including the Sun — a work begun in 1855 which was to be considerably enriched by photography from 1875 on. This was how Huggins came to demonstrate that the nebula Orion is gaseous, but that that of Andromeda, which is in fact a galaxy, is composed of stars. So astrophysics was gradually born out of spectroscopy. Cosmology, still rudimentary, was to follow a little later from the fusion of these two sciences. The hole pierced in the shutter by Newton was in fact a keyhole on the universe.

The analysis of the solar spectrum carried out by Newton has scientific and technical offspring which are numerous, if not innumerable, from optical spectroscopy to atomic and molecular spectroscopy, via microwaves, infra-red, electronics, to X-rays, radio frequencies, and many more.

A belated explanation

It was not until the 20th century that we succeeded in explaining the existence of different coloured bands in the solar spectrum, and that only after the Englishman Ernest Rutherford had demonstrated in 1911 that atoms are made up of positively charged nuclei surrounded by negatively charged electrons. It was Niels Bohr who, in 1913, established that atoms have levels of discrete energy and produce a light of a specific wavelength when electrons transfer between these levels. The type of atom in an object corresponds, therefore, to a 'colour' in the spectrum, that is to say a given band of the spectrum. The relative number of atoms corresponds to the relative intensity of the band. In fact the solar spectrum reflects not only the type of atoms making up the Sun's surface, but also that of the ions and radicals as well as the type of atom and ion in the Sun's atmosphere and in free space which are to be found between the Sun and the Earth. The latter absorb light from the Sun at their characteristic wavelengths rather than emit it, giving rise to the dark lines observed by Fraunhofer.

Newton and modern painting

One might claim that, in a way, the discovery of the light spectrum by Newton lies at the origins of the Impressionist and particularly the Divisionist schools of painting. These schools suggested the reconstitution of coloured forms in light, not by use of the fundamental colour of the objects represented, but by the juxtaposition of touches of primary colours, the artistic balancing of which helped to achieve a far greater fidelity, at least if the painting were viewed from a certain distance.

Liquefaction of gases

Faraday, 1818

Gas in its other form

Transforming gases into liquids was an idea that had occurred to noone ...

Starting as an errand boy in a bookshop, then as a bookbinder, the Englishman Michael Faraday had a passion for books which led to an intense interest in physics. This in turn led him to follow the evening lectures of the celebrated Davy, who noticed him and took him on as an assistant. At the time a great interest was being taken in gases and their characteristics: elasticity, modification due to the effect of heat, etc. Faraday tried to determine their rates of compression with the help of a machine he had invented. He compressed chlorine by lowering its temperature and was amazed to find that the gas changed into a liquid. This was in 1818 and his exploit attracted a lot of attention — so much so in fact that Davy took umbrage at it and opposed the nomination of his former assistant as a member of the Royal Society.

Helium 3

In 1971, as an experiment, helium 3, an isotope of ordinary helium 4, was liquefied and an important discovery made: at very low temperatures, liquid helium 3 loses all viscosity and when poured into a container it tends to run away laterally. Moreover, it is a superconductor, that is to say that an electric current passed through it meets no resistance and persists almost indefinitely, a property which is used in numerous researches in nuclear physics.

Liquid oxygen

The fact remains that Faraday uncovered an unknown physical phenomenon: the action of temperature and pressure on gases. In this way he liquefied hydrogen sulphide and sulphur dioxide, but failed with oxygen, hydrogen and nitrogen. It was not for lack of knowledge that Faraday failed with these gases but rather for lack of great enough pressures. Some sixty years later, the Frenchman Cailletet, a master ironmonger (an interesting detail as Faraday, his predecessor, was a blacksmith's son) invented a machine capable of producing and maintaining pressures in the order of several hundreds of atmospheres. In 1877 he managed to liquefy oxygen by the stratagem of putting the gas under a pressure of 300 atmospheres, then relieving the pressure. The resulting sudden expansion of the gas brought in its wake a devastating fall in temperature, to $-118.9°C$. He beat the Swiss Pictet to it by a mere 10 days.

Chain liquefactions

Engineers began to explore the benefits for industry of the liquefaction of gases, the fruit of Faraday's 'little experiments'. In 1895 the German Linde liquefied air by the same process as Cailletet — compression and expansion — but using an intermediate recooling. It was the Frenchman Georges Claude, however, who first invented the industrial process for manufacturing liquid gas, by achieving fractionated distillation which allowed for the isolation in liquid form of oxygen, hydrogen and argon. The Cailletet pump inspired many researchers; by 1899 the Englishman Dewar had succeeded in obtaining boiling liquid from hydrogen. Then, in 1908, the Dutchman Onnes achieved liquefaction of helium, the final gas to be liquefied for industrial uses.

Liquid crystals

Reinitzer, 1888; Lehmann, 1889

Neither mineral nor organic

*Half-crystal, half-liquid, for a long
time there was no interest shown in
them. Their true potential remains
to be discovered.*

The discovery of liquid crystals was one of
the slowest to emerge. In 1888 the Austrian
botanist Friedrich Reinitzer was observing,
to further industrial research, the reactions
of cholesterol benzoate, when he noted that
this compound had two different melting
points: at 149°C this solid became an
opaque liquid, and at 179°C it became
clear. This phenomenon intrigued several
chemists and physicists and shortly afterwards the German, Otto Lehmann, discovered that, during its opaque phase, the
salt in question showed areas of crystalline
molecular structure. It was Lehmann who
created the expression 'liquid crystals'.
These liquid crystals therefore had two
nominal discoverers, the first having determined their existence and the second their
nature.

The crystals awoke the interest of more
than one expert. More and more substances
were being discovered which, while still
being liquids, presented crystalline structures, and it was the French crystallographer, Georges Friedel, who was to

clarify matters in 1928. Liquid crystals
exhibit the peculiarity of comprising crystalline structures in the solid state which
remain until the first melting point
(observed by Reinitzer); the material is
then called mesomorphous, and possesses
certain characteristics of solids and others
of liquids. Beyond this first melting point
the crystal actually becomes a liquid and is
then said to be isomorphous.

Three kinds of crystals

It was Friedel who established the three
universally accepted groups of liquid
crystal. The first is the smectics, consisting
of one or two layers of oblong molecules
whose axis is perpendicular to the plane on
which they are laid out and which, when
heated, slide over each other without losing
their orientation or exchanging molecules.
The second group is the nematics, which
consist of only one layer and also have an
axial orientation which is perpendicular to
the plane but which are mobilized by electric or magnetic fields, passing from the
clear to the opaque state and vice versa.
Finally come the cholesterics, which are
composed of several layers of molecules
whose main axes are parallel to the plane
but slightly divergent from one layer to the
next in such a way that the axes form helical
(spiral) structures perpendicular to the
plane. Most cholesteric liquid crystals are,
as the name indicates, derived from chol-

Soap and coleoptera
The simplest of the liquid crystals can be easily
produced: these are soap bubbles, crystals of
the nematic type. The shells of certain
coleoptera (insects), reflect polarized light and
possess a rotatory power exactly comparable
to cholesteric crystals.

esterol; it was these which Reinitzer observed. This last group possesses one exceptional optical property: circular dichroism. This means that, when a light ray strikes them some wavelengths are polarized in a circular fashion and others reflected; this results in cholesteric crystals exposed to white light exhibiting iridescences characteristic of the angle of incidence of the light and of the temperature of the crystals, and these iridescences intensify and vary if the temperature of the crystals is raised. Another peculiarity of cholesteric crystals is that when they are cooled down they pass through a smectic phase.

In slow motion

So many singularities could not fail to intrigue the scholars who strove to explain, according to the laws of physics, how the same material could maintain crystalline properties at different temperatures. Among the most illustrious theoreticians to tackle the question were Louis de Broglie, W.H. Bragg and Max Born; but in vain. Most of the explanations included an assumption which was not verified, that the crystals studied contained impurities and

---

Mysterious molecules

To date we have only very imperfect knowledge of how organic crystals react inside a living system; probably the reaction depends on electric currents (nervous influx) and thermal variations. As these elements and these factors exist in number, the field of investigation for biologists in this area seems to be vast and still very largely unexplored.

---

that, that apart, it was simply a question of particular crystals. In fact the researches fell through because the crystals in question offered no practical benefit.

An assured future

The situation changed with the use of liquid crystals in the display systems of measuring instruments, including watches. The crystals used, most often of the nematic group, are contained between a reflective metal layer and a sheet of glass. The metallic layer being the bearer of circuits, induction of a current brings about major disruptions in the interior of the crystal which render it opaque and give it a definite coloration; natural light or a miniaturized lamp emphasizes the areas heated by the circuits.

Implications for biology

The application of liquid crystals in display systems and, since 1982, in television screens, represents only a tiny part of the interest in this material. It can be said in fact that liquid crystals constitute a largely incomplete discovery. These substances, the study of which belongs to organic chemistry as much as to crystallography, seem to be of interest in many areas of biology. Since the end of the 1960s it has been known that some systems of living organisms in fact constitute liquid crystals of the cholesteric type. Compounds of lecithin (a complex lipid present at different stages of metabolism), cholesterol, biliary salts and water have thus revealed crystalline structures under a polarizing microscope. Even DNA, the essential constituent of genetic inheritance, exhibits crystalline properties reminiscent of liquid crystals (see p. 53).

Lucy

Johanson, 1974

The first to walk upright

A pre-human skeleton, three and a half million years old, has decisively established the ancient origins of the human line.

On 30 November 1974, some 150 km to the north of Addis Ababa in Ethiopia, in a place called Hader, in the Afars, the American palaeoanthropologist Donald Johanson, who was excavating this region which is rich in prehistoric remains, made an un-dreamt-of discovery: an almost complete skeleton of a hominid. A hominid is a line of descent separate from the great apes and constituting a stage on the way to the *Homo* race, and a line in which individuals are characterized by distinct traits such as an upright stance.

A one-metre tall woman

Once the skeleton was reassembled it became clear from the proportions of the pelvis that it was that of a female; it was christened 'Lucy', after the Beatles' song 'Lucy in the sky with diamonds', for it was a fabulous discovery. One metre tall, endowed with a tiny skull, Lucy, according to the specific formation of her pelvis, had walked upright, a categorical sign of evolution towards hominization. The knee joint confirmed this.

Lucy's age

It remained to establish the 'age' of this fossil. The dating of mineral samples from the site by the potassium/argon method (see p. 35) gave an age of 3 million years, with

a margin of error of plus or minus 200 000 years. However, dating with palaeo-magnetism indicated an age of 3.5 million years. This method of dating is based on the fact that the Earth's magnetism has changed several times over the course of the millennia. As it has remained 'printed' in the layers of rock in formation, in molten lavas in the baked earth, and as it has been possible to reconstitute its successive variations relatively exactly, the age of a specimen can be measured by ascertaining its magnetic characteristics.

It was only in 1977, after many verifications, that the international scientific community began to take the discovery made by Johanson and his team from the International Afar Research Expedition seriously. The journal *Science*, organ of the American Association for the Advancement of Science, an unchallenged authority, agreed to publish Johanson's discovery of a hominid previously called *Australopithecus afarensis*. Johanson also published a theory of evolution of hominids up to the appearance of modern man, *Homo sapiens sapiens*, in terms of the Lucy discovery. For him Lucy represented the first known stage of the upright posture or bipedal, and therefore a link between the great apes and the hominids; however, *A. afarensis* was not a tool-maker and therefore belonged to a stage of evolution prior to that of *Homo habilis*, which appeared 2 million years ago. *A. afarensis*, which measured a little more than 1 m in height, would, said Johanson,

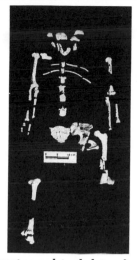

The most complete skeleton. *Lucy's skeleton was found amongst the remains of five or six individuals, including two children about five years old. It stood out as being the most complete of all.*

basing his idea on extremely complex anthropological indicators, have given birth to two family branches. The first of these was the australopitheci already discovered, *A. africanus* (also called *gracilis* because it was of short stature and poor constitution) which appeared 2.5 million years ago, and *A. robustus* with a height of 1.50 m which appeared 1.5 million years ago. The second branch consisted of *Homo habilis*, appearing 2 million years ago and measuring 1.50 m; followed by *Homo erectus*, which appeared at the same time as *A. robustus*, that is 1.5 million years ago and already approaching a height of 1.50 m; and finally *Homo sapiens*, measuring between 1.50 m and 1.80 m, which appeared 200 000 to 300 000 years ago.

The idea of two separate lines of descent was dictated by the clearly differing degrees of evolution of two specimens which appeared at the same time, *A. robustus* and *H. erectus*. Two such different contemporary types could only have come from very different strands of evolution. Furthermore, there was a clear anthropological relationship between *H. erectus* and its immediate predecessor, *H. habilis*, whilst there was considerably less of a link between *H. habilis* and *A. robustus*, although the latter had appeared later. This genealogical tree, which is hardly contested at all, at least in its main lines of descent, is quite coherent. At the origin of the species to which Lucy belonged, *A. afarensis*, Johanson placed the Ramapithecus, which is 8 million years old and of which any description must remain speculative since no complete skeleton has yet been found.

Lucy's ancestors?

The point on which the community of palaeoanthropologists, notably the famous Leakey family — Louis, the father; Mary, his wife; and Richard, the son — was to differ from Johanson is this one: Johanson held that the movement towards humanization began with Lucy, that is to say with *A. afarensis*. As far as he was concerned, Lucy was the distant grandmother of humanity. It is certainly true that no earlier bipedal specimen has been found. We have in fact no definite knowledge of whether the ramapiethsecus moved on two limbs or four.

Bipedalism

For the American Owen Lovejoy, locomotion on the hind legs does not, properly speaking, constitute an evolution from ape to man, for the ape can walk very well on its hind legs and only uses its hands when adapting to a forest environment. According to this specialist in locomotion, bipedalism is linked to a reproduction strategy which consists of having few offspring and so taking very great care of them. Thus the female would have taken to standing upright in order to carry them in a safer way. This method of locomotion could well have appeared in the savannah lands.

Lystrosaurus

Anon, 1969

The burning poles

The discovery in Antarctica of the fossil of an animal which also lived in Africa verifies the theory of continental drift.

In 1969 a Canadian mission discovered in Antarctica the fossilized remains of an animal which lived at the end of the primary era, that is to say in the Carboniferous and Permian periods. It belonged to a group of reptiles, the Synapsida, who in a sense 'prepared' for the arrival of the mammals, certain of whose characteristics they already exhibited. It was a lystrosaurus, of the therapsida subgroup (the other group being that of the pelycosaurians). In its reconstituted form it is reminiscent of the present-day hippopotamus.

The animal was already known to palaeontologists: it lived some 220 million years ago, but in Africa. It was formal proof of the drifting apart of the continents, a theory advanced in the 1920s by the German, Wegener, and which was no longer in any real doubt but the mechanics of which remained to be determined.

According to this theory, the Earth consisted, at the beginning of the primary era, of a single continent, Pangaea, surrounded by water. Gradually, over the millennia, Pangaea, formed from light silicates or 'sial', broke up and the fragments drifted on the layer of heavy silicates of the Earth's crust at the bottom of the oceans, the 'sima'.

Originally greeted with much scepticism, Wegener's theory continued to win supporters, for the evidence in its favour continued to pile up. In the 1950s came a more refined theory which took over. This was the theory of plate tectonics, which stated that the continents were in fact resting on plates of the Earth's crust which in turn floated on a sea of magma. It was the continental drift which formed, for example, the great mountain ranges through folding in the Earth's crust.

A climatic indicator

According to plate tectonics Africa and Antarctica were formerly joined together. The discovery of the lystrosaurus formally confirmed this point, just as it also brought confirmation that the climate of Antarctica and that of different regions of the Earth has changed a lot through the ages. The variations in the Earth's rotational axis, the cooling of the globe, and also the changes brought about by the new mechanics of evaporations and precipitations have been some of the factors of climatic change. The oceanographic soundings of the last 30 years have confirmed and refined the new theory.

Lystrosaurus *This distant ancestor of the hippopotamus was by no means one of the most 'interesting' animals of the Carboniferous and Permian periods, but the presence of its remains in Africa and Antarctica at the same time confirms that these two continents were formerly a single one.*

Magnetic field induced by electric current

Ørsted, 1820

Electromagnetic induction

*Every electrical current creates a
magnetic field. This property was
discovered by accident.*

One evening in April 1820, the Danish
chemist and physicist, Hans Christian
Ørsted (also written Oersted), holder of the
chair of physics at Copenhagen University,
gave a lecture on electricity. He passed a
current from a battery through a wire held
by a wooden support to demonstrate, it is
thought, an aspect of Poisson's theory on
potential. Quite by chance, near to the electric wire stood a compass. When the current
was passed, the needle of the compass swung
round and pointed away from North. The
induction of a magnetic field by electricity
had just been discovered. It was to have
considerable repercussions in the following
months. Ampère, Faraday, Poggendorff,
Ohm, Maxwell and many others took up
the idea, each one working out fundamental
applications. Ohm, for example, was to use
the magnetic field to measure the intensity
of electric currents; Faraday was to discover
electromagnetic induction and the influence of dielectrics on electrostatic fields.

The magnetic moment

Ampère further explored electromagnetic
induction as follows: taking two electric
wires along which he passed a current in
the same direction he noted that the wires
were attracted to each other, but that if he
reversed the current in one of the wires it
repelled the other. If the experiment was
done with two small electric coils, the results
were similar and the forces exerted between
them were equal to those of two magnets;
one coil could be replaced by a magnet

without altering these forces. One could
also determine the magnetic moment,
related to the force experienced by a magnet
in a magnetic field, by the dimensions of
the coil, the number of its loops and the
current circulating through it. In any case
the complex law governing the force exerted
in such circuits was established in the very
year of its discovery by two Frenchmen,
Jean-Baptiste Biot and Félix Savart, and it
is still this same law which is in use today.

The significance of Ørsted's discovery is
immense. On the practical level it has led,
for example, to electromagnetic induction
(see p. 60), and to the discovery of Hertzian waves. On the theoretical level it has
led, after a brief diversion, to the transition
from the Newtonian concept of universal
laws to the relativistic concept. The links
between electricity and magnetism have in
fact, since the end of the 19th century,
caused physicists to admit that a single
mechanism could not adequately explain
all the relationships between all the
elements.

***A needle under
influence*** *In Ørsted's
experiment, which can
be easily reproduced, the
needle of a compass placed
near to a wire or coil with
an electric current will align
itself with the magnetic field
engendered by the current.*

Memory (Molecule of)

Braestrup, 1980; Chapouthier, Rossier, and Dodd, 1983

Towards the memory pill

*A chemical substance which would
stimulate the memory is an old
dream of neurologists.*

Research into the possibility of a molecule capable of stimulating the memory goes back to the early decades of the 20th century. It was successively believed that the 'mnesical' or 'mnemogenic' substance could be found in cocaine, phosphoric acid and calcium. In 1957 the Englishmen Steinberg and Summerfield of London University carried out the first serious work on the subject by demonstrating, inversely, that certain substances such as nitrous oxide (the 'laughing gas' which was an early anaesthetic (see p. 16)) could impair the memory.

In 1971 the Italian Ungar proposed the existence in the brain of nucleic acids whose consumption by other individuals would allow them to acquire the memories of the brain which had recorded them in the first place. His experiments on worms and rats were compromised by the counter-experiments carried out by other specialists. The same year, the American Pauling postulated that memory depended on traces of elements in the brain, a theory which remains incomplete. In reality it is extremely difficult to intervene chemically in the memory processes, which to this day are unknown from the neurological point of view.

In 1980, the Dane Braestrup tried to find traces of a hypothetical natural tranquillizer in human urine, similar to endorphins (see p. 65). After treating 1700 litres of it he isolated a molecule of the family of β-carbolines, alkaloids found in various plants like the fruit of the passionflower, whose structure changes their effects, making some stimulants and some sedatives. However, the β-CCE isolated by Braestrup exhibited one particular feature which did seem to indicate that the researcher had achieved his goal, since it bonded itself to the nerve receptors of tranquillizers.

Tranquillizer or convulsant

Counter-experiments undermined Braestrup's conclusions: β-CCE was not a tranquillizer but a convulsant, and furthermore, the molecule in question was not natural, but artificially formed at the time of treating the urine by the reaction of certain substances with organic compounds.

Braestrup's experiment, however negative it may have been, did demonstrate one useful fact: the affinity of β-CCE for certain cerebral receptors. In addition its convulsant effect recalled the series of trials set up by Steinberg and Summerfield, considerably enriched by the tranquillizer experiment, on the relationships between memory and psychotropes (substances capable of affecting mental activity): convulsants stimulate the memory, tranquillizers impair it. It is therefore possible that β-CCE activates the memory process.

In order to verify this hypothesis, the Frenchmen Chapouthier and Rossier and the Englishman Dodd, of the National Centre for Scientific Research, tested in 1981, β-CCM, a molecule very close to β-CCE, also a synthetic molecule.

Experiments were carried out on mice.

Three groups of animals were put in conflict situations which were stressful or anxious-making, and subjected to a short learning test. One group was given tranquillizers, another β-CCM, and the third group nothing at all. The results showed that the animals which had best memorized their experience were those which were under the influence of β-CCM, and those which had retained the least were the animals on tranquillizers; the group which had been given nothing scored somewhere between the two.

A similar experiment carried out on humans in 1984 by the Englishman Dorow, with another molecule close to β-CCM, showed that this induced a state of acute anxiety. The effects of the β-carbolines on human memory had not been objectively measured in 1986, but other trials on animals confirmed their positive effects on memory capacity.

Distress and memory

It was established in 1984–5 that these molecules are fixed by the same cerebral receptors which pick up benzodiazepines, or tranquillizers. It follows that it is on these same cell sites that the brain traps the substances which inhibit or stimulate the memory. In 1986 a hypothetical 'molecule of distress' was still being sought, a natural molecule which would also stimulate the memory. As for the natural 'tranquillizing' molecule which would allow painful memories to be wiped out, this might be an endorphin (see p. 65).

The pharmacological use of β-carbolines as 'memory pills' was not envisaged in 1986 as something for the immediate future, because of its anxiety-producing properties.

The actual nature of memory remains myterious. It is difficult to separate it from the general activity of the cerebral cortex, that is the circuits which unite hundreds of billions of neurones through the intermediary of chemical messengers, that is actual thought mechanisms. The pathology of the brain and notably the study of lesions, effects of drugs (like LSD) and psychiatric troubles (like schizophrenia), indicate, in fact, that deterioration of thought processes is almost always linked to that of memory.

In 1966 the writer Henri Michaux reported that images evoked under LSD intoxication and therefore drawn from memory, are extremely difficult to memorize. It follows from this that every action in the memory brings into play the whole activity of the cerebral cortex and vice versa. As the discovery of Chapouthier and Rossier in particular indicates, a prompt action on the memory seems to set in motion mechanisms which are expressed by anxiety or by the state of cerebral awareness which is most favourable to memorization.

A scheme of relations

The discovery of a 'memory molecule' does not change the fact that we still know practically nothing about memory. If there is reason to believe that it resides in one region of the brain rather than any other, and if certain causes of its deterioration and improvement are known, we are still at the stage of the theory put forward in 1920 by the celebrated Spanish neurologist Santiago Ramón y Cajal and refined in 1940 by the Canadian Donald Hebb: it consists of a set of relationships between nerve cells where data are stored. This elementary point, the method of fixing data in a cell, is the fundamental key to memory, and itself remains open to speculation.

The regions of the memory

It seems that certain regions of the brain are more particularly implicated in the memorizing process, in a way which is still unknown: these are the corpus callosum and the Papez circuit.

Microwave oven

Le Baron Spencer, 1946

A pocket oven

A melted sweet reveals the thermal effect of short electromagnetic waves.

Electromagnetic waves have been known since their discovery by Hertz in 1887. Microwaves are electromagnetic waves which are found within the frequency band between about 10^9 and 10^{12} cycles per second, that is to say from one gigahertz to one terahertz, with a wavelength of between one mm and one metre; they have been known to exist since the beginning of the century. Until the middle of the 20th century their principal application was in radar. While working on waves of 25–120 mm in 1945 an American, Percy Le Baron Spencer, who was a physics engineer at Raytheon, one of the world leaders in radar equipment, noticed that a sweet which he had in his pocket had melted. He quickly realized the cause: the microwaves trigger a molecular vibration within the material they pass through and this vibration brings with it a release of heat. Follow Spencer's discovery, Raytheon developed a cooking programme for microwaves and patented the first cooking apparatus of this type, the Radarange. This machine had a power of 1600 watts. It was heavy, awkward and expensive and was originally intended for use in hospitals and military canteens. In 1967 the Amana company, a subsidiary of Raytheon, put the first household microwave oven on the market.

Magnetron

A device for generating microwaves, it comprises an evacuated chamber with a central cathode surrounded by a circular anode. Electrons expelled from the cathode circle it under the influence of electric and magnetic fields. Microwave production results from resonances between the moving electrons and cavities in the anode. Developed during the 1940s for radar, it is widely used in microwave ovens today.

A dangerous experiment

It must be noted that the incident which led Le Baron Spencer to discover the thermal effect of microwaves could have been dangerous, for at certain frequencies microwaves can prove fatal.

Mutations

Dubinine, 1937; Spencer, 1957

The creation of the human species?

---

*The possibility of hereditary genetic
variations is one of the major
questions of biology.*

In the 18th century when the first ideas of the evolution of species were being formed, several scholars, including Buffon and Cuvier, the founder of palaeontology, began to suspect, while studying the first exhumed fossils, that there was a logic in the sequence of the species through the geological ages. So Cuvier found that the head of the *Palaeotherium* strongly resembled that of the contemporary tapir. He did not however go as far as transformism (the theory of evolution), which held that a species would end up, after a series of transformations, as a related but different species. He was hostile to transformism, especially to the theory advanced by Lamarck. Nevertheless, the question of the mutability of species had been raised, and in spite of Cuvier's efforts to consign it to oblivion, it would remain an issue for some time; certainly well beyond the publication of Charles Darwin's works in the 19th century. Despite violent hostilities, Darwinism, a kind of transformism refined by principles of adaptation and the natural selection of species, was to take root.

However, there was no proper scientific basis for Darwinism during the early decades of the debate about it; this theory, founded on rigorous observation of minor variations of different species, of which the birds of the Galapagos Islands are a striking illustration, rested only on deduction and reasoning according to the following outline: by virtue of small mutations, one species is changed into another, because it has adapted itself to a specific environment whereas another has not been able to do so;

this latter species is therefore eliminated by natural selection, for reasons of climate, food availability, disasters and so on. Discussion about the mutability of species tends to degenerate into speculations about the 'jumps' made by nature in order to get from one species to another, as maintained, for example, by Geoffroy Saint-Hilaire a few decades earlier, or even about progressive transitions upheld by Gaudry among others. No distinction was made between strictly adaptive changes which concern only the phenotype (the expression of the genetic capital as influenced by the environment) and intrinsically genetic changes (the genotype); the very good reason for this was that genetics had not yet been born (see p. 90).

Germen and soma

At the beginning of the 20th century two precise concepts were formed: that of the genotype, called germen, and that of the phenotype, which came to be known as the soma. Until then biologists, who, with good reason, had not yet started on molecular biology, had been reduced to observation of phenotypes or *soma*, found themselves in a situation rather like that of someone trying to get an impression of a tapestry by examining the reverse side. They knew nothing of the infinite variety of genetic formulae in living beings. This ignorance would only gradually be shed. Their intellectual habits were not always favourable towards new ideas. So it was that when the

Russian Dubinine discovered, in 1937, that in each population of fruitflies (subjects favoured by geneticists because the generations succeed each other rapidly and the chromosomes are easily observed), at least 2 per cent of individual flies were found to be spontaneous mutants, few biologists paid any attention. The discovery was certainly a major one, but there was as yet no molecular biology to explain it; biologists, too attached to the old idea of the stability of species, preferred to consider the mutations discovered by Dubinine as 'accidents' of no great general interest. As Jacques Ruffié noted in his *Traité du vivant*, it was not until 20 years later that the American Spencer struck a fatal blow against the 'traditional concept of mutation, considered as an exceptional element'.

Darwinism

After having carried out series of cross-breedings, Spencer discovered that each fruitfly harboured in its genetic capital at least one mutation which had a visible effect. Darwinism (which had meanwhile become neo-Darwinism) would seem to have had its first scientific verification; not so, for the neo-Darwinians held that mutations must have an effect tending to uniformity. Since 1966 the experiments of Harris, Lewontin, Hubby, Stone, Johnson, Kanapi and others (experiments carried out with the technique of electrophoresis) have proved that, in all species, the presence of mutations is a regular phenomenon, seen in all populations, even those subjected to severe selective — or Darwinian — pressures.

These discoveries are both Darwinian and anti-Darwinian; in the first place they show that there is a constant potential in all living species for hereditary variations; but in the second place they exclude the hypothesis of general variations which modify the whole species, at least in constant circumstances. The 'catastrophism' of certain neo-Darwinian biologists, which suddenly proclaimed totally new and uniform species, was discredited; so indeed was the 'fixism' dear to creationists, who were inspired by the Bible and according to whom species did not change.

In 1977, gathering statistics on the amount of variability or genetic polymorphism in living species, the Frenchman Lucotte discovered that the rate decreased as one went up the scale of living species: it was therefore lower in the lower vertebrates than in invertebrates, lower in the higher vertebrates than in the lower, and lower in man than in the higher vertebrates. The reason for this seems to be structural: the more capacity for adaptation a being has, the less it needs genetic variation. Thus the human brain will have reduced man's genetic variability. There we have a third fundamental discovery, since it establishes a direct relationship between intellectual development and biological outcome. It seems that the fact of man's variability has been well demonstrated, always excepting any monstrous anomalies caused by genetic accidents. Since 1977, however, a new theory has come to light, thanks to a discovery by the Japanese, Tonegawa; it has been taken up by others and has caused something of a sensation.

Why are there no human mutants?

In spite of the fact that a certain number of natural mutations have been produced in the fruitfly, the species is none the less stable. That is explained by the very weak reproductive capacity of mutants as much as by the capacity of the genetic capital to protect itself. The mechanisms of self-protection get stronger as one goes up the scale of the species. No viable mutation in man is known and the two per cent rate of natural mutations in the fruitfly is absent from the human species. That is the reason why there are no human 'mutants'. The genetic capital is expressed in such a way that the mutants, those whose DNA is different, are either incapable of developing or of reproducing themselves.

Darwinism and mutations

One of the difficulties met by biologists in the study of the evolution of the species bears on the fact that all the mutations would have been fully viable; this is theoretically possible but hardly seems to be supported by present-day knowledge of biology.

A disturbing fact

Since the discovery of DNA and its double helix structure by Crick and Watson (see p. 52), geneticists have maintained a practically mechanistic idea of Mendelian genetics (see p. 129); each individual would simply be a reflection of the combined genetic capital of his father and mother, each of these being responsible for half. DNA was considered, in a way, to be a living Meccano set. Then Tonegawa discovered a strange fact, disturbing, almost absurd: in the genes there was a lot more DNA than was needed for genetic reproduction. For example: while 10 units or nucleotides were sufficient for one gene, it was found to have 100. Only 10 were used for genetic expression, or were active in governing the synthesis of such and such a substance; these are called exons. There were 90 left over, however; these are called introns and seemed to serve no useful purpose.

This was soon proved wrong: in 1980 it was realized that these extra introns were not in fact inactive. They intervened in order to repair damaged fragments of DNA. The Frenchman Slominski also discovered that they could undergo mutations: mutations which were not gratuitous, for they could, for example, block the synthesis of certain proteins.

To the disgust of certain geneticists attached to a sort of 'fixism' which appealed to the fundamental immutability of the specific germen of the human race, everyone was forced, from the 1970s, to allow a certain margin of mutation in certain groups of human populations. This explains the fact that the β molecular chain which enters into the composition of human haemoglobin (the oxygen-carrying component of blood) varies slightly according to ethnic groups. 'This chain', wrote Ruffié, 'is a peptide composed of 146 amino acids arranged end to end according to a fixed sequence.' (A peptide is a chain of amino acids, which are the building blocks of protein.) The sixth acid from end is normally glutamic acid, but there are variants. One is known in which valine acid takes the place of glutamic acid; this produces a different haemoglobin, called S, and is responsible for a type of anaemia frequent in African populations, called sickle cell anaemia, and for drepanocytosis, a serious illness which is an adapation of the malaria which is endemic in those regions. Yet another variant is known where another amino acid, lysine, occupies the sixth place mentioned above. In this case we are dealing with a less serious form of disease, characterized by a slight anaemia and a different type of haemoglobin called C. If the same lysine were to replace glutamic acid but in the 26th position on the chain, then haemoglobin E would be obtained, a characteristic of certain South-East Asian populations and yet another known variant. Several variants of the β chain of haemoglobin have been indexed and it is thought that mutations of the same type might well affect many other peptides.

So in the 1970s the notion took shape that the DNA of all living creatures 'recombines'. (One might even say this is a 'selfish' substance which only reproduces and reconstitutes itself for its own benefit.) This recombination can take place in the individual, but it can also take place in the species, or more exactly, in groups of isolated populations of a species; this has been seen in the matter of β chain of haemoglobin, leading to a new species definition.

In spite of its experimental proofs this idea was not unanimously accepted for it evokes a doctrine, rejected more than a century before, called Lamarckism.

Lamarckism

One of the fundamental theories of Lamarckism was the inheritance of acquired characteristics; it did in fact, seem that the transmission of a variant in the β chain of haemoglobin, an acquired variant, constituted an argument in favour of Lamarckism. This was just an illusion, Lamarck having claimed that modification of an organ by function could be transmissible, an idea as scientifically unsound as saying that the children of a lumberjack would have large biceps because their father had developed his in his work. Evidently Lamarck knew nothing of genetics. If the theory of mutations through the inheritance of acquired characteristics is upheld in the 20th century, it is on an exclusively genetic basis.

Rebel germs

One of the most interesting examples of the transmission of acquired characteristics was the fruit of a collective observation, and therefore an anonymous discovery, made throughout the world in the 1960s: numerous germs, originally very susceptible to antibiotics, had suddenly become 95 per cent resistant to them. Whereas in the 1950s 10 000 units of penicillin were enough to cure an infectious endocarditis, in the 1960s 50 000 units were required. Again, in the 1970s, the appearance in Asia of a new family of penicillin-resistant gonococci was observed, and one has lost count of the species of bacteria against which it has been necessary to mobilize an entirely new arsenal of antibiotics. Research helped to establish rapidly that the resistant bacteria had learned to secrete enzymes which neutralized antibiotics: penicillinase, for example, which rendered penicillin ineffective.

A dark chapter

Trofim Lysenko, Director of the Institute of Genetics at the Soviet Academy of Sciences from 1940 to 1965, did his best to impose Lamarckist notions on his country's geneticists, particularly in the agricultural field. According to these notions it would be possible to modify plant species by grafting, arguing for an imaginary circulation of 'plastic material' between the graft donor and receiver. In short, Lysenko wanted to achieve by force mutations by the transmission of acquired characteristics, designed to show the superiority of the Marxist ideology of intervention in nature. These naive efforts would have been of little importance if Lysenko had not reduced a number of his adversaries to silence, indeed had them deported, causing a serious setback to Soviet genetics and writing one of the darkest chapters in the history of this discipline.

At first it was believed that abuse of antibiotics had created a species of more resistant bacteria by natural selection, killing off the weakest and letting the more resistant ones survive, but after several false starts such as this, it was realized that the new resistance of the bacteria was due to a true genetic mutation. Certain bacteria had, by a mechanism we are still striving to explain, produced a gene which synthesized an enzyme antagonistic to the antibiotic; and, quite remarkably, the bacteria exchanged this information, allowing this gene to be produced not only within the one species, but also from one species to another. It is supposed that this gene would be made up of introns of bacterial DNA (see p. 81).

Bacterial mutations

Bacterial mutations were not, strictly speaking, an absolute novelty: in 1926 the American Griffith had succeeded in producing controlled mutations of pneumococci, and it was already known that certain characteristics of a strain of microbes (for example, its capacity to produce certain enzymes or secrete certain antigens) appeared when the subjects of this strain multiplied in contact with the corpses of another strain. However, we did not even begin to penetrate this mechanism until the arrival of molecular biology. It is now known that the transmission of information is carried out by a 'pocket' of bacteria, called the plasmid, a genetic element carrying genetic information and able to exist outside the nucleus of the bacterium.

The stability of these mutations is not certain and is currently a matter for debate. It is a difficult question since the lower species are, as has been noted above, much more open to mutation than the higher vertebrates.

In thousands of years . . .

Attempts have been made, however, to produce mutations in vertebrates; they have not really been successful, for the basic genetic mutations which have been achieved do not seem so far to be transmissible, or else, it is thought, they are accompanied by pathological (harmful), indeed fatal mutations which disappear with the death of the subject. It is true, though, that genetic engineering is a completely new science and that we are still far from unlocking all the secrets of DNA. The stable mutations which have been obtained in vertebrates (for example, in domestic animals) are defined as secondary. For example there has as yet been no success in bringing about a mutation in the sow which would increase its secretion of milk. It is also true that apparently minimal mutations take a very long time: some 20 million years were required for the emergence in the *Mastodontes trilophodontes* (a type of mammoth) of a third molar. This means that one or two human generations represent a derisory period in which to observe induced mutations.

Our ancestors the apes

The uncertainty about how stable mutations come about has not much compromised the theory of definitive mutations in higher vertebrates. From around 1970 the French biologists Lejeune, Grouchy and Ruffié have postulated that the branch of the great anthropomorphic (human-like) apes in Africa which appeared in the middle of the Miocene era would have been caused accidentally in the following way. The apes changed from 48 to 47 chromosomes by translocation (rearrangement) or displacement of a chromosome which passed in one lump into a gamete, (reproductive cell) then by changing from 47 to 46 chromosomes, as modern man has. In fact the comparative study of the karyotypes (arrangement of genetic information) of the African great apes and of man has helped to establish the chromosomes responsible for this modification; genetically we differ from the great apes by barely two per cent.

The cause of this major mutation, which ended in the creation of the human species, is not known, but oncology (medical study of cancer) has identified three possible causes: radiation, especially the ionizing type, chemical products, and viruses. It has now been established beyond doubt that

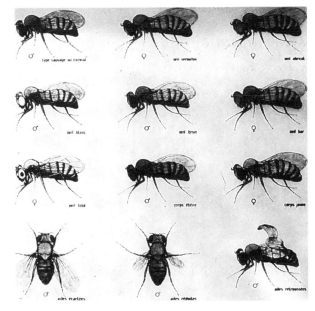

Some types of spontaneous mutations observed in the fruitfly

Favourite subjects for the study of genetic variations by virtue of the rapidity with which generations succeed each other, and the size of the chromosomes.

ionizing radiation induces chromosomal changes (see p. 68). It is not currently known if these changes are hereditary, although there is, in the specific case of cancer, a presumed hereditary transmission of 'expressive' oncogenes (cancer-triggering genes; see p. 141). It may be supposed that in a population group which is isolated and subjected to constant ionizing radiation, there might have been a sufficient number of individuals in whom the same genetic mutations were produced, and whose cross-breeding would have culminated in the formation of a new species.

Permanent mutants

Difficult to demonstrate in higher animals, and probably lethal in the majority of cases, mutations are however currently 'practised' by the beings which are to be found in the lowest echelons of life, notably by viruses. It is known that numerous viruses, including those of flu and AIDS, have an amazing capacity for adaptation. The flu virus has available several types of genome (genetic make-up) which allow it to vary from year to year, a little as if a higher being had four or five different physiological entities at its disposal from which he could choose according to the prevailing circumstances.

Hypotheses

This ionizing radiation could have been natural: a kind of natural nuclear reactor has been found in Africa, due to a strong presence of uranium in the soil. It might also have been accidental, caused by the fall of a meteorite, for example. The same meteorite might also have scattered in the environment chemical substances likely to affect the genes. Finally the hypothesis of viral infections having modified the genotype might equally be true.

'Chance and necessity'

These are just theories, however, and many geneticists estimate that if such causes have been able to influence mutations at one particular time, they cannot be systematically evoked to explain the thousands of mutations which have taken place since the single-cell organisms which lie at the origin of living species on Earth. It seems more probable that the mutations which have led to species changes have been the product of genetic recombinations (see p. 81) which take place according to the laws of chance and necessity (to echo the famous expression of Jacob and Monod).

Natural radioactivity

Becquerel, 1896; P. and M. Curie, 1898

The radiation which surrounds us

The spontaneous emission of energy
by matter was an unexpected bonus
of the discovery of X-rays.

A few months after the discovery of X-rays by the German Röntgen (see p. 223) in 1895, it was known that the production of these rays in a vacuum within a tube was accompanied by strong phosphorescence in the glass of the tube. Some researchers wondered, intrigued by this unknown phenomenon, if certain substances rendered phosphorescent by visible light, like phosphor itself, would emit a comparable radiation. This idea had emanated from the celebrated mathematician, Henri Poincaré. It was then that Henri Becquerel, who came from a famous scholarly family, had the idea of testing it. Becquerel was interested in a rare ore, uranium, the salts of which he had studied. He enclosed some potassium sulphate of uranium, among other ores, in black paper which he placed on a photographic plate. He discovered that this element produced a weak image on the gelatine of the plate. Becquerel repeated his experiment and verified that the observed effect was independent of phosphorescence as it was known, and that the uranium salts, kept in the dark for a long time, retained their properties.

These observations were published by the scientific journals. One evening Marie Curie, pupil and wife of Pierre Curie, found the account of Becquerel's observation and pointed it out to her husband. The mysterious radiation described by Becquerel intrigued them both. What was its nature and where did its energy come from? Therein lay a fine thesis for a doctorate, for no-one had yet tackled the subject.

Marie Curie's doctorate

After many attempts and several months of research, Pierre Curie obtained the use of premises where he could further examine the phenomenon. It was a shed in Paris, on the ground level of the Ecole de Physique, rue Lhomond, which had been used until then as a store. Marie Curie, who was a doctoral candidate, began by testing a secondary phenomenon described by Becquerel: the mysterious radiation had in common with X-rays the fact that it took the charge from an electroscope, so making the surrounding air conductive. The radiation was therefore of the type that caused the formation of ions, ie ionizing radiation. Marie Curie decided to measure this ionization by using two instruments invented by Pierre Curie and his brother Jacques: the Curie electrometer and the piezoelectric quartz. She first noted that the intensity of radiation was proportional to the quantity of uranium oxide studied, and that this radiation was independent of the conditions of observation.

Pierre Curie

When the Germans Walkhoff and Giesel declared in 1900 that radium, discovered by Marie Curie, had effects on organisms, Pierre Curie exposed his arm to it. A lesion appeared which he described himself: 'The skin became red over an area of six square centimetres.'

Marie Curie then had the idea that the phenomenon — which she was the first to call radioactivity — owed its origin to atomic reactions, since its intensity was proportional to the degree of uranium in the element studied. In the course of her investigations she drew on the mineral collection of the Ecole de Physique and discovered in 1898 that another element, thorium, produced almost the same effects as uranium. Radioactivity was established as a generic phenomenon. If the original discoverer of radioactivity was Becquerel, it was Marie Curie who discovered its atomic nature.

The causes of radioactivity

Once on her way, Marie Curie would discover two new radioactive elements, polonium and radium (see p. 157). It remained for the scientific world to analyse the atomic causes of radioactivity, which would be accomplished in stages.

In 1899 Marie Curie discovered that a body exposed to a radioactive source itself became radioactive, and that this secondary radioactivity decreased in time, but at a more rapid rate than the primary radioactivity of the source. The basics for the theory of radioactive decay were laid down but the principle itself would not be formulated for a few more years. In 1900 the Englishman Ernest Rutherford noted that bodies rendered radioactive behaved as if they were covered in a thin radioactive layer which could be eliminated by scraping or with the help of powerful acids, but that the radioactive matter destroyed by the

acid, for example, turned up again in equal quantities after evaporation of the acid. A platinum wire rendered radioactive thus ceased to be so when made white hot; but the radioactive molecules became scattered over cooler surrounding surfaces. Rutherford also made a still more important discovery; that the radioactivity given off could be fixed to a certain degree on the cathode of an intense electric field, which showed that it was positively charged.

Alpha, beta and gamma radiation

In Canada, where he had become director of a laboratory, Rutherford discovered that radioactivity is not homogenous but made up of three types of radiation. The first, easily absorbed, can be stopped by a sheet of paper or a few centimetres of air; it is the weakest type, which he called alpha. The second is much more penetrating, and can cross an aluminium partition several millimetres thick; he called this beta. The third can pass through a block of iron 20 cm thick or a lead plate several centimetres thick: this is gamma radiation. Alpha radiation consists of atoms of helium whose range varies from 2.5 to 8.5 cm according to the emitting body and their speed is some 1600 km per second. Beta radiation consists of a flow of electrons ejected at a speed approaching that of light in extreme cases. Finally, gamma radiation consists of electromagnetic waves of variable energy.

Radioactive decay

It was Rutherford who, in 1903, set out the general theory of radioactive decay, which would be refined by subsequent generations of physicists. The atoms of radioactive substances are unstable; at given, regular intervals of time, often infinitesimal, a set number of these atoms decay spontaneously and release alpha and beta particles.

In 1905 the German Schweidler put forward the notion that the number of atoms decaying in a given time remains more or less constant for a given number of atoms; this would later be verified by Geiger, Kohlrausch, Meyer, Regener and others.

Marie Curie

From 1897 onwards Marie Curie handled radioactive ores without any precautions, being unaware of the dangers. Thirty years later she would feel the first effects of her involuntary imprudence: she was affected by a double cataract and mysterious fevers reminiscent of tuberculosis. In fact Marie Curie was suffering from aplastic anaemia, which seriously affected her bone marrow and from which she died in 1934.

$$E = mc^2$$

After a certain length of time a radioactive material ends up losing part of its mass and therefore part of its radioactive intensity (the number of atoms per unit mass undergoing decay remaining constant.) It was just after 1910 that it was decided to give the name 'half-life' to the time at the end of which a radioactive substance loses half its mass. Half-lives vary considerably depending on the radioactive element itself: that of radium, called radium 226, is 1620 years; that of radium 218, its isotope, is 3.18 minutes.

The obvious connection between the difference in mass of the element and the products of its disintegration (m) and the energy released by its decay (E) was to dictate to Albert Einstein in 1905 the famous formula $E = mc^2$, where c represents the speed of light. This formula lies at the origin of attempts to exploit atomic energy, the first application of which was the invention of the atomic bomb in 1940.

Radioactive elements and sites

From the time of the discovery of radioactivity in 1896 it was established that there existed in nature elements which decay spontaneously. It was not until several years later, and then only gradually, that scientists became aware of the dangers of radioactivity. In several parts of the world scientists measured levels of radioactivity in various substances. From the beginning of the century it was accepted that a fairly low level of natural radioactivity existed in the environment, due to the presence of radioactive impurities in ordinary elements; this is how granite comes to release radium gas or radon. One gram of ordinary metal contains infinitesimal traces of radio-elements in the order of 10^{-15}. In 1907, however, the Englishmen Campbell and Wood discovered that elements not classed as radioactive, such as potassium and rubidium, gave off radioactivity, not because of traces of radio-elements but because they themselves were the seat of disintegration. By 1950 four others had been discovered: indium, lanthanum, samarium and lutetium.

It was in the 1960s that systematic measurements of the levels of natural radioactivity in soils began to be taken. From this it emerged that France is one of the countries possessing the greatest number of radioactive sites. In the southern region of Hérault near Lodève, for example, is the site comprising the greatest amount of natural radioactivity in the world: it releases more than 1.7 sieverts per year, which is more than 30 times the maximum permitted dose for workers in the French atomic industry and 300 times the permitted dose for the general public. The cause of this is a stratum of uranium from the hill of Riviéral which releases 0.21 millisieverts per hour; this means that anyone spending 24 hours there receives the maximum allowed dose for a year.

Everyday radioactivity

Ecological preoccupations have also led to explorations of other possible sources of natural radioactivity and it has been found that, besides granite which at times releases significant levels of radon, red brick and concrete also give off a certain amount of radioactivity, more or less comparable to that of granite (from 2 to 4 millisieverts a year).

Drinking water can also be radioactive. So it is that with 180×10^{-11} curie per litre, Vichy-Grande-Grille is one of the most radioactive types of water in France, closely followed by that from the spring La Savoureuse at Ballon d'Alsace (640–

The danger of meteorites
Recently fallen meteorites ought to be handled with caution for they very often contain radio-elements such as argon 37 whose half-life is 35 days.

Radio-elements in the high atmosphere
Several radio-elements are formed high in the atmosphere under the effect of cosmic rays. As well as carbon 14, there is hydrogen 3 or tritium, beryllium 7 and beryllium 10, one of the isotopes with the longest half-life at 2 700 000 years.

1060×10^{-10} curie per litre) and that of Monteil Bonnac-La Côte (1800×10^{-10}). These last two, however, are not commercially produced.

Another source of natural radioactivity is the constant bombardment of the Earth by cosmic rays, which increase tenfold in spring and which, especially in the high atmosphere, raise to some 0.03 millisieverts the dose received by passengers on transatlantic flights.

Medical radiography

From the 1970s on, medical officials of numerous countries have asked doctors and administrators to limit the number of X-rays; they were worried because of the amount of radioactivity given off by them and added to that already in the natural environment and to the emanations of artificial radioactivity (accidents occurring at power stations, fallout from experimental explosions). A typical dose from a chest X-ray would be 0.02 millisieverts and the same from a dental X-ray. Scans involve higher doses: a liver scan 1 millisievert, the same for a lung scan, but 5 millisieverts for a bone scan. To put these in perspective, the natural annual background dose for the UK population is 2.3 millisieverts per head; the maximum dose permitted to the general population from man-made sources is 1 millisievert. The maximum permitted doses vary from country to country and that they are generally based on empirical valuations, for the possible effects of low radiation have not yet been determined exactly.

The dawning of awareness of the sources and risks of radioactivity has therefore been slow. This collective 'discovery' has been made more than half a century after Becquerel's lucky breakthrough.

Neptune

Le Verrier, Galle and Adams, 1846

A mysterious disturbance

Almost simultaneously and quite separately, three astronomers discovered this planet in our solar system.

The planet Uranus was discovered in 1781 by Herschel (see p. 208), and its orbit and expected position on the sky were precisely calculated; but in 1845 its position was troubling the astronomers: it showed a divergence of two arc-minutes in 20 years from the expected position. This could only be due to a large mass whose gravitational force of attraction was disturbing the movement of Uranus. In October 1845 the young English astronomer Adams had correctly calculated the orbit of this unknown body; but his superior, Airy, had no confidence in him. The same year, the Frenchman Le Verrier undertook his own calculations and on 31 August 1846 he published them. On 23 September Le Verrier asked his German colleague Galle to look for the planet at the given point in the sky; the latter did so and that very evening discovered Neptune.

The unbelievers

The English realized, after Galle's discovery, that not only had Adams been correct, but also that they had been provided with three accidental observations of the planet which they had ignored. ...

136

Nervous reaction (Chemical transmission of)

Loewi, Dale, 1936

All action is electrochemical

Nervous reactions are brought about by the relay of chemical substances called mediators.

In the 1930s there was no doubt at all concerning the electrical nature of nervous impulse, which had been discovered many years before by Galvani and which had, since that time, given rise to much research. Since the time of the German Waldeyer (1891) it was known that the neurone is the unit of the nervous system and it had been possible to measure the rapidity of nervous impulse which varies, moreover, according to the nerves. However, certain reactions which followed the impulse and seemed to bring chemical substances into play were not satisfactorily explained. The balance between sympathetic and parasympathetic nervous systems (the two of which tend to act in opposite ways to each other) was not a sufficient explanation.

Difficult to identify

It is likely that there exist many other neuromediators apart from those recorded so far. However, their research and identification pose technical and scientific problems which are often considerable. In order to decide whether a substance really is a neuromediator, it has to be present in the presynaptic part of the nerve to a significant degree; its effect has to be identified by external injection of the supposed active substance; and there have to be enzymes specific to it which are inhibited by theoretically antagonistic drugs.

Acetylcholine

Among those addressing the question were the German Otto Loewi and the Englishman Henry Hallett Dale. In 1919 Dale had isolated acetylcholine, a chemical substance which, amongst other effects, slows the rate of the heart. The mechanism for the production of acetylcholine still remained obscure even after this discovery. Loewi had, in the course of his research from 1921–6, become increasingly convinced that there was an essential link between nervous impulse and the release of substances such as acetylcholine, but he had neither proof nor knowledge of the method of secretion.

Loewi was based at Graz in Austria. A professor at that city's university, he was, he said, inspired in a dream to take a particular route in order to confirm his suspicions. He stimulated the heart nerve endings of the parasympathetic system of a frog and by this method slowed down the animal's heart rate. Then, with the help of an extracorporal circulation circuit he introduced the blood of this frog into the body of another frog which had not been stimulated. The heart rate of the second frog also slowed down. This was the proof that the stimulation of the parasympathetic system in the first frog had triggered the secretion of an active substance. On analysis this was revealed to be acetylcholine, the substance isolated by Dale. In 1936 Loewi shared his Nobel prize for physiology and medicine with Dale.

Loewi's discovery (and Loewi's alone in this case) gave rise to the chemical theory of nervous transmission. According to this, the nervous impulse triggers, at the extremity of the nervous fibre, the release of a chemical substance called a neuromediator. The example of acetylcholine shows, to a certain extent, the soundness of this theory, for it has been established that acetylcholine is released at the joining of the parasympathetic system and the muscles (the neuromuscular junction). The release of the neuromediator at the neuromuscular junction modifies the electrical potential of what is called the muscular plate, which governs the tightening and relaxing of the fibres. The intensity of the effect depends on the duration of the nervous impulse or, more precisely, on the number of electrical impulses; a certain number of synchronized impulses of all the related nerves is necessary in order to cause the release of the neuromediator. This chemical system also includes substances antagonistic to the mediators: these are enzymes, also released at the same neuromuscular junction. By destroying the neuromediators after a few milliseconds these enzymes ensure that the action will be short-lived. In this way cholinesterase destroys acetylcholine.

Loewi's discovery improved our understanding of a certain number of phenomena which had previously been obscure, such as the action of certain drugs and toxins. For instance, we now know how the botulin toxin causes paralysis in the person poisoned: by the blockage of the neuromuscular junction, which prevents the release of acetylcholine and cause the heartbeat to accelerate. It has also been possible, in the light of this discovery, to invent drugs which, for example, correct the action of the destructive enzymes if it is either too slow or too fast.

The neuromediators themselves are found in the form of molecules in small sacs or vesicles situated on the nervous fibre at the presynaptic nerve terminals — that is to say, in front of the synapse or point where the two fibres meet.

The work carried out in the 1960s by the Englishmen Hodgkin and Katz made it possible to establish that in the nervous tissues themselves, the action of the nervous impulse also occurs in a chemical manner: by the entry of sodium ions and the exit of potassium ions in the case of stimulation, and inversely in the opposite case.

Chemical mediators

The whole of neurophysiology and physiology has benefitted considerably from this discovery, which cast new light on the chemical balances of internal systems and the consequences of their disturbance in certain diseases.

In addition to acetylcholine numerous other chemical mediators have been discovered since 1936, and certain groups have been distinguished. These include the biogenic amines, adrenaline, noradrenaline, serotonin, dopamine and histamine; the amino acids, either inhibitors or stimulators; and the polypeptides, like the enkephalins and the endorphins (see p. 65), vasopressin, etc. During the last half-century, it has been established that these substances play an extremely wide range of roles, in areas such as growth (somatostatin), childbirth (oxytocine), Parkinson's disease (dopamine) and hypertension (adrenaline). Finally, it is known that certain neuromediators, like endorphins, are hormones and that, furthermore, hormones (the steroid type) can act as neuromediators or neuromodulators on the neurone. It is in this field, which is much wider than had been suspected, that endocrinology meets neurophysiology, although the hormones act on the neurone outside the synapse and therefore in a way that is not specifically neurological.

An important application

One of the most celebrated applications of Loewi's discovery has been the invention of tranquillizers, drugs which counteract the action of neuromediators.

Nitrogen

Rutherford, Priestley, Cavendish, and Scheele, 1772; Lavoisier, 1777;
Chaptal, 1790

78% of the air's volume

Discovered almost simultaneously by four great researchers, this element was only adequately described by Lavoisier.

At the end of the 18th century scientists were striving to make an inventory of the world's natural substances and the fundamental mechanisms of nature. Without doubt air was one of the first substances which captured their attention. What is it made of? Partly of oxygen — this had been known since 1772 thanks to the Swede Carl Wilhelm Scheele — but what else? Many researchers suspected that in addition to 'inflammable air', the name then given to oxygen, there was another gas.

The same year, 1772, a Scottish botanist, David Rutherford, had the idea of burning air in a closed container (by placing a candle under a bell-jar) and testing what was left, for there was gas remaining in the container. Rutherford pumped this gas through lime and noted that it was unable to support combustion of a candle. He called it 'phlogisticated air', that is 'burnt'; others would call it 'dirty air'. Also in 1772 the English clergyman and chemist Joseph Priestley carried out a similar experiment resulting in the same conclusions, as did a second English chemist, Henry Cavendish, and a Swedish pharmacist, Carl Wilhelm Scheele, who had discovered oxygen.

Most minds of the time were burdened with fairly philosophical theories on 'phlogistics'. At that time it was thought that an element called phlogiston (which does not, in fact, exist) separated from every combustible body at the time of its burning.

Lavoisier

In 1777, in a paper presented to the Academy of Sciences, Lavoisier returned to the phenomenon of combustion. He demonstrated that this was not due to the liberation of a hypothetical 'phlogiston' but to the combination of a body with oxygen at a given temperature. He recognized the gas discovered by his predecessors as an independent gas which he himself named azote (nitrogen) from the Greek word *zoon*, meaning 'life', preceded by the negative prefix *a*, because this gas is, according to Lavoisier, incapable of supporting life. In the 19th century, it was found that numerous plants in fact survive only by the nitrogen in the air (see p. 140).

After Lavoisier, it was the famous French chemist Jean-Antoine Claude Chaptal who discovered that 'azote' was a constituent of saltpetre, or potassium nitrate. In 1790 he named it nitrogen (chemical symbol N).

A cycle of life

Nitrogen is an essential constituent of all living matter and its cycle in nature constitutes one of the principal cycles of life. A small part of the nitrogen that we breathe is dissolved in the blood and plays a direct part in the building of our proteins.

In cosmology

Spectral analysis of molecules present in the form of clouds between the stars has enabled us to detect nitrogen which is, with hydrogen and carbon, one of the most abundant elements in space. Hence the idea that there might exist, elsewhere than on Earth, 'organic' molecules, whether organized or not, especially in areas of abundant hydrogen and favourable chemical and thermal conditions.

Nitrogen (Fixation of)

Boussingault, 1855; Hellriegel and Willforth, 1887; Winogradsky, 1891; Haber and Bosch, 1908

A vital cycle

Understanding the role of nitrogen in plant life was a laborious process, but eventually this element's cycle in nature was established.

Every plant contains nitrogen, an essential constituent of amino acids. Where does it come from? From the air which is full of it or from the soil which also contains it in different forms? At the beginning of the 19th century there was simply no answer to this fundamental question of general physiology.

The first to try to answer it was the Frenchman Jean-Baptiste Boussingault. At first he could only obtain contradictory results: a series of experiments in 1837–8 revealed that clover and peas grown in artificial soil contained more nitrogen than the original seed but that there was no excess nitrogen in corn or barley. From 1860–71 he carried out a second series of experiments: he placed earth, the mineral composition of which he had measured, in sealed containers without air. After a period of time, if the total quantity of nitrogen had not varied, the quantity of nitric nitrogen had itself increased, that is to say there had been oxidation in the soil of ammonified nitrogen. Schloesing and Müntz, pupils of Boussingault, confirmed in 1877 that nitrification, or the formation of nitrogen salts, was a biological phenomenon. The agent of this process was still to be determined.

One important hurdle was cleared in 1866 by the Russian, Woronine, who discovered that the nodules on vegetable roots were full of bacteria; it was evident that it was by their mediation that certain plants were nitrogen-fixing. It was at the end of another series of experiments that Hellriegel and Willforth discovered in 1886–7 that in fact vegetables assimilate free nitrogen from the air thanks to their root nodules. In 1891 the Russian bacteriologist Serge Winogradsky, who would later establish himself in France, was studying soil and discovered in it the bacteria responsible for producing nitrates or nitrites. He also discovered bacteria which equally produce nitrates on strictly mineral soils, which implies that they fix nitrogen from the air (the first type fix it from the soil).

Very many researchers were involved in trying to understand better the mechanism of nitrogen fixation: Schloesing again in 1874, Müntz in 1889, Mazé in 1898, and Molliard in 1909. All stated that the higher plants could absorb nitrogen just as well in mineral form as in the form of nitrates or ammonia, provided that the concentrations were weak.

A new discovery by Winogradsky, two years after the first was a bacterium, *Clostridium pasteurianum*, which could not live in the presence of oxygen, but could in the presence of nitrogen, which it then fixed in the rarefied air of the earth.

> **Industrial process for nitrogen fixation**
> At the beginning of the 20th century there was concern about the growing need for nitrogenous fertilizer to produce the increased amounts of food necessary for the growing world population. Fritz Haber, a German organic chemist, showed how nitrogen from the atmosphere could be used to make ammonia, a discovery for which he was awarded the Nobel prize in 1918. With the industrialist Robert Bosch he established the Haber-Bosch process, still used today to make most nitrogenous fertilizers.

Oncogenes

Weinberg, Cooper and Wigler, 1981

The first of the keys

The specific genes which can cause cancer have greatly contributed to the understanding of the subject.

In the years before 1981, one aspect, among others, of the formation of cancers intrigued specialists; the question was as follows. A cancer, whether its cause be viral, ionizing or chemical, takes several years to develop. During this time the tissues have renewed their cells several times. Normally the cancer cells would have been eliminated in the course of these renewals. So why did they perpetuate themselves? Could it be because they might originate outside the mechanism of cell renewal, in the genes for example?

The fragment of DNA responsible

At the end of the 1970s, three teams of American researchers — Professor Robert Weinberg at the Centre for Cancer Research, a branch of the Massachusetts Institute of Technology; Professor Geoffrey Cooper of the Sydney Farber Cancer Institute of Boston; and Professor Michael Wigler of the Cold Springs Laboratory of Long Island — separately undertook to verify the genetic hypothesis according to

the same experimental procedure: the total DNA of cancerous human cells was taken and grafted on to the cells of mice (non-differentiated cells or fibroblasts, because such cells tolerate grafts without rejection). The fibroblasts were cultivated in the laboratory. Those which had absorbed the human DNA and become cancerous were isolated; a sample of the DNA was again taken, but this time broken down with the help of restriction enzymes (see p. 187), and the fragments were again grafted on the fibroblasts. Some of these became cancerous; others did not. Those which did were again isolated, their DNA extracted and fragmented, and the operation begun again a certain number of times. At the end of this series of grafts it became evident that by a process of elimination one was getting closer and closer to isolating the segment of human DNA responsible for cancer. The result of all this was that each of the three teams discovered that there were indeed fragments of human DNA which were responsible for cancers.

Cancer genes

These cancers were not all the same: Cooper's team isolated the breast cancer gene, Weinberg's that of cancer of the colon, and Wigler's that of cancer of the bladder which also seemed to be responsible for lung cancer and cancer of the colon.

Unless we suppose that every human being is born with a carcinogenic gene or oncogene which will manifest itself sooner

---

Nucleotide

A nucleotide is a compound of a nucleoxide and phosphoric acid, which forms the principle constituent of nucleic acid. Nucleic acid is the general term for natural polymers in which bases are attached to a sugar phosphate backbone.

or later, it must be the case that the responsible genes are abnormal or become abnormal. How and why? Without answers to these questions the discoveries of oncogenes achieved in 1981 remain incomplete.

An uncontrolled protein

An oncogene was isolated and analyzed nucleotide by nucleotide, in order to establish its structure; it was the oncogene of cancer of the bladder, and comprised 930 bases, termed, as usual, A, C, G and T (see p. 54).

It seemed quite normal, except with this difference: at point 653 instead of finding a base T (thymine), a base G (guanine) was found. The normal combination of T–G had been inverted to become G–T, and because of this the proteins synthesized by this gene were abnormal. Furthermore, the regulation of production of this protein had been knocked out, owing to the solidarity among genes. It was about to go into uncontrolled reproduction. This is certainly not the only factor working in cancerization, but it is one of the most important mechanisms which has been brought to light.

From gene to oncogene

The oncogenes are not situated haphazardly on the chromosomes, and not all genes can become oncogenes; there are particular genes which assume the role of triggers. This doubtless concerns regulatory genes and not structural genes.

At first it seemed that a gene changes into an oncogene under the influence of one of the three factors mentioned above, or of a combination of two or three of these factors, even though it be hereditary. The change also occurred when there was a failure of the repair mechanisms of DNA, which seemed to act in such circumstances.

A trigger for cancer

Having become an oncogene, the gene in question did not appear to manifest itself immediately, but would remain for some time indeterminable and potentially car-

cinogenic. Its manifestation, which could signal the unleashing of cancer, appears to depend on a supplementary factor which could be of a statistical nature. So it is probably that a single oncogene in a cell is not sufficient to trigger the cancer, and that this will not appear without the presence of a given number of identical oncogenes in several cells.

A hereditary imbalance

It is very probable that other external factors play a part in the automatic suppression of cancerous lines, and conversely in their survival and proliferation, notably the action of the immune system; several works have considerably reinforced this theory.

The discovery of oncogenes cannot be presented as the only key to cancer, but it is an essential aspect of a succession of imbalances and anomalies leading to the disease.

A viral origin

It is certain that at least some oncogenes are of viral origin. The proof of this is that a virus gene has been identified which is almost identical to a natural growth gene, and that cancer cells are characterized by a mode of growth which it is impossible to control by normal regulatory mechanisms. The proliferation of cancer is therefore explained by the existence of this strange growth gene, which does not respond to the autonomous commands of the cell.

Oncogenes and 'rogue' fragments

It seems that one can, hypothetically at least, compare the oncogenes of viroids, protoviruses or rogue fragments of DNA. In effect there is a suspicion among biologists that the formation of oncogenes could be due to the insertion of a foreign fragment of DNA, such as a viroid, which would not even originally be infectious; this would give it the capacity for foiling the immune system. If this hypothesis were proven, it would be necessary to add protoviral causes to the recognized viral causes of certain cancers (very limited in number).

Ovulation
(Substances acting against)

Haberlandt, 1921; Allen and Corner, 1929; Marker, 1941

A magic root

The invention of oral contraceptives resulted from a series of lucky accidents.

At the beginning of the 20th century contraception was a taboo subject, and its practice was limited in the Western world to the use of coitus interruptus, observations of the woman's rhythms of fertility, and the use of mechanical contraceptives. Knowledge of the genital systems and their physiology, especially the female system, was still far from complete.

A singular experiment

It was with a certain degree of surprise that the scientific world received the discovery of the German Haberlandt in 1921. Starting from the still hazy idea that pregnancy triggers the secretion of anti-ovulatory substances, this physiologist completed a singular experiment: he transplanted ovaries from pregnant rats and rabbits to non-pregnant females and obtained in the latter an experimental sterility. A little later he discovered that by injecting the animals, indeed by simply letting them consume ovarian extracts from pregnant females, he achieved the same results. Haberlandt postulated that these were specific hormones secreted only during pregnancy and which brought about sterility; he coined the expression 'hormonal sterilization' and suggested that there could now be a move to experimentation on humans. He did not then realize what questions would be raised by this. In the 1920s concern for repopulating countries devastated by war made the idea seem completely inappropriate, and anyone undertaking it risked a prison sentence.

Progesterone

Some researchers, however, continued to address the matter and, in 1929, the Americans Allen and Corner succeeded in extracting and crystallizing a hormone extracted from the corpus luteum (part of the ovary) of the sow. Their discovery was interesting because it shed light on the hormone involved, but had no practical application, for in order to obtain a few milligrams of it, Allen and Corner had to deal with the carcasses of thousands of sows. It was only in 1934 that the hormone was isolated in a steroid; it was called theelin in the United States, progynon in Germany, oestrine in Britain and folliculin in France.

Ancient methods of contraception
Recourse to plant steroids for contraceptive ends preceded the birth of 'the pill' by a long way. In ancient times rue, savin and pennyroyal mint were used. In Africa albizia, which interrupts the development of the egg in 12 hours, no matter what stage it is at, has been used since time immemorial. An elementary molecule, mimosine, which reduces fertility can also be found in several species of mimosa.

Allen and Corner had suggested calling it progesterone. It was not considered as a contraceptive, but simply as a possible remedy for glandular problems, and that in the distant future. It was not then known how to synthesize this hormone, and even in 1934, research into, and clinical application and descriptions of contraconception methods fell foul of the anti-obscenity laws in many countries. (The first birth control advisory clinic, opened in 1916 by the American Margaret Sanger in the United States, was closed by the police, and Margaret Sanger was sent to prison for one month.)

For the American Russell Marker, the barriers to contraception would fall when the hormone in question became available in large quantities, so he followed a scent laid by the chemists of the US Bureau of Plant Industry, in 1941. This organization, a branch of the Federal Bureau of Agronomy, had studied the reputed contraceptive properties of infusions used by the Indian women of Nevada. Many primitive tribes throughout the world have for centuries used various plants to sterilize women and control population. Experiments indicated that infusions of at least one plant, *Lithospermum ruderale*, had contraceptive effects on mice. Pursuing research into such a controversial subject in the United States was risky, and so Marker left for Mexico, persuaded that another plant, wild yam, possessed the same properties. Marker gathered considerable quantities of this tuber in the Mexican jungle and, amazingly, discovered in it progesterone.

It seemed that Marker had benefited both from local folk lore and from very acute intuition. What is certain is that he succeeded in synthesizing over the space of three years 2 kg of progesterone, a prodigious quantity at that time. The story goes that he went into the offices of a small Mexican pharmaceutical firm, asked the manager if he was interested in the industrial production of progesterone, and then put down four jars on the manager's desk, each containing a pound of the hormone.

Waiting for 1960 ...

Marker's discovery, the third in the history of oral contraception, stayed on ice for nearly 10 years; no-one dared to experiment with progesterone on a woman, partly because it was not yet known what mechanism it might inhibit or activate. It was the research of the American Gregor Pincus which finally brought the discovery into the clinical domain. Pincus, a biologist in the making, entered a large New York birth control centre where, ascertaining the ineffectiveness of traditional methods, he began progesterone experiments on rats. In 1956, he undertook three series of clinical trials, one among the women of Port-au-Prince in Haiti, and the two others among Puerto Ricans. The trials were conclusive and in 1956 the firm of Searle of Chicago, to which Pincus was an adviser, commercially produced the first known contraceptive pill, Enovid. It remained alone in the market for a decade; the federal authorities only granted a licence to Enovid for the treatment of gynaecological troubles and not as a contraceptive. It was only in 1960 that the real aim of the pill was acknowledged.

Oxygen

Scheele, 1772; Priestley, 1774; Lavoisier, 1775-7

A 'modern' element

Hazy theories at first obscured the discovery of this element.

A discovery can be made by more than one person; that of oxygen would be no different from many others, were it not that, of its three successive discoverers, only the last really gave it both its full meaning and its name.

A tortuous journey

Physics in the 18th century was completely tangled up in a theory of nature much more philosophical than scientific: phlogistics, which can be summed up in the belief in a 'flammable' element in nature which renders objects inflammable. So, when the German-speaking Swede Scheele, one of the most brilliant experimenters of his time, discovered oxygen, he hardly guessed its nature. Scheele, who was studying gases in the years 1768-70, enclosing them in bladders to analyze them, did not discover oxygen while analyzing the air, as might have been expected. His journey of discovery was much more complex, not to say unexpected and tortuous. He observed that, when he heated nitric acid to a high degree he obtained brown smoke which, collected in a balloon flask containing lime water, only partly dissolved; an odourless gas remained. He then developed an explanation which nowadays appears most odd. The brown smoke was analogous to that

which was produced when nitric acid was heated in the presence of metals like manganese, that is to say, according to the popular theory, when nitric acid had drawn out the 'phlogiston' from the heat to which it had been submitted. It followed from this that the odourless gas obtained must have been an element of heat.

The 'air of fire'

This gas burned whilst giving off heat and producing a lively flame, which meant, for Scheele, that it was the 'air of fire' or 'flammable', which combined with the phlogiston to form heat. He believed he could confirm his theory by heating other bodies, like saltpetre or potassium nitrate, which also gave off the same odourless gas, whose composition Scheele did not know.

Priestley's air

It has been formally established that Scheele discovered oxygen two years before the Englishman Priestley discovered it in his turn. Did Priestley get wind of Scheele's discovery? The point cannot be finally settled, for while it is true that Scheele's work was not published until 1777, it is known that the physicist Bergman took account of it in 1774, three months before Priestley began his own experiments. It is possible that Priestley was intrigued by Scheele's discovery and wanted to take it up in order to verify it; it is also possible that he had been intending to carry out this experiment for a long time.

From liquid to solid

Oxygen becomes liquid at $-183°C$ and then exhibits a pale blue coloration. At $-218°C$ it becomes solid.

What is known is that on 1 August 1774, Priestley obtained the same gas by strongly heating one of the materials used by Scheele, residues of oxidized mercury: he deduced from this that the gas obtained was 'dephlogisticated' air. Paradoxically Priestley published his discovery three years before Scheele, that is to say, in 1774, the very year of his experiment.

The notions of modern chemistry were still a long way off. It was the Frenchman Lavoisier who, returning to the question between 1775 and 1777, clarified the nature of the gas discovered. He began from the claim, inspired by the experiment, that there existed a certain gas which had the tendency to form acids by combining with different substances. It was from this that he gave the gas its name, oxygen, derived from the Greek words *oxys*, meaning 'acid', and *genos*, meaning 'production'. For him oxygen engendered acidity and his theory was reflected in the names given to oxygen in other languages, for example the German *Sauerstoff* ('acid material'). Certainly this reasoning was not absolutely accurate, as later discoveries showed, notably when the Englishman Davy established that hydrochloric acid did not contain oxygen (1815). However, it was of immense merit in clarifying ideas.

The combustion of air

In 1771 Lavoisier was interested in the role of air in combustion, including that of sulphur and phosphorus. He had noted a weight increase after combustion which he attributed to the combination of air with the products consumed. Until then this weight increase had been explained according to notions of phlogistics, by the fact that fire penetrated the metal at the time of combustion. Lavoisier then decided to measure the variations in weight very precisely in order to see what conclusions might be drawn from them. So he carried out combustion in sealed containers and he concluded that there was absolutely no variation in weight when a metal changed into lime under the effect of heat. The theory of phlogistics was being undermined at last.

After combustion a gas remained (Lavoisier called it 'mofette') which was evidently nitrogen, and an oxide; Lavoisier heated the oxide, from which he derived metal and oxygen. The oxygen fed the flame of a candle. Lastly, Lavoisier added the 'mofette' to the oxygen and reconstituted the original air. It was thereby proved that the metal's increase in weight was due to its combining with oxygen.

The notion of elements

This experiment constituted the turning point of physics and marked a move away from the old chemistry, still close to alchemy, towards modern chemistry. Its importance went a long way beyond the simple discovery of oxygen, since it eliminated the notion of phlogistics, an imponderable element, which immeasurably complicated and even falsified explanations of chemical reactions. It also introduced the idea of measurement, Lavoisier having made constant use of the balance, as the only guarantee of the value of possible explanations of reactions.

On a more mundane level, the discovery of oxygen at last explained what happened — what always happens — when metals are heated. According to phlogistics, the metal was separated from the phlogiston; in reality, the metal unites with oxygen and so is oxidized.

The discovery of oxygen also completely modified the previously held notion of an element. For the phlogisticists, only four elements existed: fire (or phlogiston), earth,

---

The laws of sympathy

The decline of phlogistics, invoked until then to account for chemical phenomena, did not go smoothly, not only because the scholars were attached to their ideas, but also because they again found themselves incapable of explaining why one body reacted with another. By Newtonian attraction between the particles? By reason of a redistribution of the phlogiston during reactions? Because of a mysterious law of sympathy, derived from Democritus? It was necessary to await the arrival of mechanical chemistry before this question could be answered.

air and water. From this point on only a body which represented the final point of analysis and could not be further decomposed would have a right to be called an element.

Lastly, this discovery led Lavoisier on to the discovery of respiration, which was to be just as decisive (see p. 184).

The 'antiphlogistic' theory

It is therefore right to consider Lavoisier as the real discoverer of oxygen, for it was he who gave it its meaning and assigned it a place in chemical reactions which was almost exact (with the exception of the acids); neither Scheele nor Priestley, for all their achievements, had done this.

The publication of Lavoisier's work, in 1772, did not attract the assent of the scholars, who were far too attached to the 'magic' theory of phlogistics. Thus, the Englishman Cavendish refused to incorporate Lavoisier's discovery into the old way of thinking. Starting from the principle that hydrogen was water charged with phlogiston, in other words that water was in the gas, he calculated that it would require two units of hydrogen for one of oxygen if one wanted to reconstitute water,

all of which manages to be exactly right and completely wrong at the same time. That demonstrated for Cavendish the accuracy of the phlogistic theory, since the combustion of the two, that is to say the expulsion of the phlogiston, produced water. That was tantamount to saying that in the end it all came back to water, an element.

Around 1785, however, chemists began to show more and more interest in Lavoisier's theory, which was, at first, called 'antiphlogistic'.

A new nomenclature

The rational chemical nomenclature introduced by Lavoisier, founded on the new notion of what an element was and published in 1789 under the title of 'An elementary treatise on chemistry', won over the whole community of chemists during the 1790s. It enabled them, as it had enabled Lavoisier, to introduce the notion of chemistry into the study of biological phenomena, like the production of animal heat.

Lavoisier was not to enjoy his triumph for long. He was executed in 1794 after the French Revolution because 'the Republic does not need scholars . . .'

Penicillin

Fleming, 1928

A therapeutic revolution

*This half-discovery of Fleming's
was only completed 10 years later by
Chain and Florey, who finally
brought about the industrial
production of antibiotics.*

The Briton Alexander Fleming was already 48 years old when, in 1928, studying cultures of *Staphylococcus aureus*, the bacteria responsible for the formation of pus, he noted a singular phenomenon: one of the Petri dishes, glass saucers used in laboratories all over the world, had been accidentally contaminated by a green mould, and the culture of staphylococcus had virtually disappeared. He identified the mould: it was a type of lesser fungus, *Penicillium notatum*. He analyzed it and noted that the mould, when placed in contact with the staphylococcus, produced a bactericidal fluid, which he was the first to name penicillin.

Fleming realized the potential of the bactericidal power of penicillin; he tried to extract it for therapeutic uses but did not manage to do so. For years penicillin would only be used in bacteriological research, for example to isolate bacteria not sensitive to penicillin. Was this truly a discovery? Not entirely. In 1897 the Frenchman Duchesne had demonstrated the bacteriostatic power of certain moulds (that is, which inhibit the growth and reproduction of bacteria). Fleming's discovery might have remained at the same stage as Duchesne's, if it had not been taken up by two other reseachers.

The first of these was Howard Florey, a pathologist who had discovered that saliva and other types of mucus contained bactericidal enzymes; while researching other bactericidal substances he became interested in penicillin. In 1939 he joined forces with Ernst Chain, a physicist and chemist, and both rapidly succeeded in what Fleming had been unable to achieve: isolating and purifying penicillin. Their idea was to crystallize penicillin by cooling it, in order to obtain it in stable form. A Britain at war, however, did not offer them the opportunity to continue their work. Confident of their experimental work, Florey left for the United States in 1941 and there managed to persuade the government to undertake mass production of the drug. Things progressed at lightning speed; the Americans first arrived at a method of cultivating penicillin in bulk in large tanks instead of doing it in thousands of flasks, and then discovered other branches of the penicillin family. In 1943 the crystallization by cooling of the culture juices was perfected; penicillin emerged on the therapeutic scene, destined at first for the war-

2000 penicillins

The chemical synthesis of penicillin is possible but costly; the natural method is complicated. In 1959 it became possible to isolate and analyze the active nucleus of penicillin, the acid 6-aminopenicillamic, which has facilitated the production of semi-synthetic penicillin. These now number more than 2000 types.

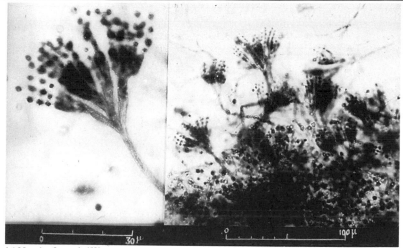

A 'fibre' of penicillin

'Efflorescences' of penicillin observed with a microscope. (The one on the left measures 30 thousandths of a millimetre, the growth on the right 100 thousandths of a millimetre.)

time battlefronts and then for the general pharmaceutical market.

Further progress was to follow: in 1951, the culture of penicillin in depth and no longer only on the surface, which considerably increased the yield of the culture containers; and, in 1959, the industrial preparation of semi-synthetic penicillins.

In 1945 Florey and Chain shared with Fleming the Nobel prize for medicine.

A very ancient discovery

The identification of penicillin by Fleming by no means marks the absolute date of the discovery of antibiotics; in fact 2500 years ago the Chinese had established that soya cream on which mould had developed was effective against skin infections and they were still using it. Even 3500 years ago the Egyptians were readily consuming large quantities of natural streptomycin, as analysis of skeletons of the time show. It has also been reported that, during the Second World War, Russian prisoners in German camps who were accustomed to eating mouldy bread, suffered less from furunculosis (boils) than the others.

Penicillin was the first antibiotic to be derived from fungi, but not the first of the modern antibiotics. Precedence must go to the derivations of sulphonic acids, called sulphonamides, the first of which was protosil, commercially produced in 1932. The second antibiotic discovered after penicillin was streptomycin (1943, Waksman); the third was chloromycetin (1947, Burkholder); the fourth the tetracyclines (1948, Duggar, Finlay); and the fifth erythromycin (1952, McGuire).

Extracted from urine

In the months following the Liberation of France after the Second World War, only the American troops stationed there were regularly provided with penicillin. This was sold on the black market and, in response to such a high demand, certain doctors collected the urine of soldiers who were taking antibiotics in order to extract from it, more or less legally but in any case effectively, excreted penicillin. ...

Photography

Wedgwood, 1802; Niepce, 1826

It could have been discovered at the time of the Caesars

The fixing of an image on a prepared mount, with the aid of light, hovers on the border between invention and discovery.

Few technical achievements are, in their essence, as ambiguous as photography, for from the outset it hovered between distinct discovery and technical invention.

The camera obscura

One of photography's essential components, the inverted and reduced reproduction of an image passing through a convergent lens, predates by far photography proper: this is the *camera obscura*, which the Italians had been using since the 16th century at least. It is a 'dark chamber' with a hole pierced in one of its walls; the light rays which pass through this opening form an inverted and reduced image on the opposite wall. By sticking a sheet of paper on to the opposite wall one can faithfully trace the image, a procedure used by numerous artists and described in detail in 1569 by the Neapolitan physicist Giambattista della Porta in his *Magia naturalis*.

The first apparatus with a lens

It seems that at the end of the 16th century the process was made easier by the Venetian Aniello Barbaro, a professor at the University of Padua, who fitted the opening with a convergent lens in order to improve the clarity of the image. So it can be seen that the necessary apparatus pre-existed

photography as such by about a century and a half. All that remained was to fix the image naturally without resorting to the pencil.

Silver nitrate

The first fruits of the chemical creation of the sensitive mount appeared accidentally at the beginning of the 18th century, when the German Johann Heinrich Schulze discovered that a flask of silver salts mixed with chalk and nitric acid had taken on a purple colouring where it had been exposed to light, the rest of the contents having remained white. Schulze repeated the experiment, covering one part of the flask with a stick-on label: the protected part stayed white. The phenomenon intrigued Scheele, the discoverer of oxygen, who studied it and stated that the bands of the solar spectrum were not oxydizing the silver nitrate at the same speeds.

Problems of communication

Scientific information at that time did not spread as widely or as rapidly as it does today. The observations of Schulze and Scheele were not widely reported.

In 1800 the Englishman Thomas Wedgwood, son of Josiah, the famous producer of ceramics, tried his hand at photography, since he placed a screen of silver nitrate at the end of his *camera obscura* in order to take copies more easily of the ancient scenes which decorated the products of the family firm. In fact an article on his efforts appeared in 1802 in the *Journal of the Royal Institution*, signed by none other than the famous Humphrey Davy, discoverer, among others, of anaesthesia and numerous simple elements (see p. 16). The annoying thing is that Wedgwood did not succeed in fixing his images, which became very faint. (Not a single one has survived.) However, it can be maintained that, theoretically, Wedgwood was indeed the first discoverer-inventor of photography: inventor, for he started from known basic principles; discoverer, for he did not know, as was proved, what he would arrive at.

Niepce and the Judean bitumen

A quarter of a century later, it was paradoxically the lack of artistic talent of an amateur lithographer, Nicéphore Niepce,

---

Daguerre

In January 1826, the same year in which Niepce achieved the first photograph, a painter specializing in producing dioramas (three-dimensionel scenes), Louis Jacques Mandé Daguerre, wrote to Niepce to inform him that he was trying to achieve permanent images with a *camera obscura*. Niepce reacted, as might be supposed, with circumspection and it took another three years before the two men finally became partners. Ten years later they were selling the rights of their invention – heliography, as Niepce called it – to the French government, which is therefore owner of the rights of the process, in theory at least. ....

Fox Talbot

In 1833, knowing nothing of the work of Niepce and Daguerre, an English mathematician, William Fox Talbot, in turn invented photography with a *camera obscura*. He called his invention 'photogenic drawing'; it is now called calotype. He first manufactured negatives, and having found the method of fixing them, placed them against blank sensitive paper, thus obtaining real photographs by inversion. The fact that Talbot's first photographs go back only as far as 1835 certainly denies him any claim to be the historical father of photography, but does not diminish his worth. Wedgwood, Niepce, Daguerre and Talbot all brought chance and intuition together.

---

which finally led to photography. Niepce was interested in lithography (making prints using an original etched in stone), which had just been invented; he transcribed on stone his son's drawings, by a process which took him to the brink of photography. He used silver chloride, which went black when exposed to light, in order to obtain negatives of the drawings. He was not a gifted artist and it was his son who did the drawings for him, but when his son left for the war, Niepce found that he had to produce drawings himself. He looked for a mount to put at the bottom of a *camera obscura*, just as Wedgwood had done, and he also chose paper sensitive to silver salts: silver chloride in this case, since he was already familiar with this product. Apparently he obtained better results than Wedgwood, but even when retouched with acid, his first 'photos' were not to his satisfaction. He then resorted, by instinct, to a substance widely used by the academic painters of the time, Judean (or Syrian) bitumen, drawn from the ancient Lake Asphaltite in Judea. It is a thick hydrocarbon which, when mixed with blacks and dark browns, gives a fine paste (which never totally dries). However, exposed to light it hardens, a characteristic which must have drawn it to Niepce's attention. He manufactured an original product by dissolving bitumen in oil of lavender and coating a sheet of copper with it.

The first photograph

The first 'photo' consisted of exposing this rudimentary sensitive sheet to the sun, under an outline drawing on transparent paper. The surfaces exposed to the light hardened; the others did not. When plunged into a solvent therefore, this sheet gave a negative, revealing the metal of the sheet beneath these outlines. However, by blackening these outlines the effect was reversed, since the outlines became darker than the rest of the bitumen. By introducing a similar sheet into his *camera obscura* and aiming his 'lens' at the window of his house at Gras, near Chalon-sur-Saône, Niepce finally obtained in 1826 (after eight hours' exposure) a photograph on metal of the barn. Photography had been born.

Photosynthesis

Priestley, 1772; Ingen Housz, 1779; Senebier, 1782; Saussure, 1802; Sachs, 1864

Solid light

The synthesizing of nutritious elements in plants using light only came to be understood in stages.

Like many discoveries, that of photosynthesis was made before the idea of it was really formed.

Priestley's experiments

In 1771, the Englishman Priestley, preoccupied with the theories of the time on phlogistics and the nature of air (see 'Oxygen', p. 145), began a series of experiments which did not in fact, have anything much to do with photosynthesis. He was working on what he called the 'goodness' of the air: that is to say, its useful role in respiration. To this end he put a mouse under a belljar and noted that its respiration reduced the total volume of air by a fifth, then he put a plant under the same belljar and discovered that the volume of air was restored. His measurements were quantitative only, since he would not discover oxygen for another three years, and yet more time was to pass before Lavoisier laid the foundations for the chemical analysis of air according to modern methods.

Respiration in plants

The importance of Priestley's technical innovation was considerable. In the area of what is properly called photosynthesis, his principal contribution lay in the following discovery: that plants produced what he called 'dephlogisticated air', in reality oxygen, and that this phenomenon only occurred in the presence of light. So he established the basic idea of photosynthesis.

In 1779 the Dutchman, Ingen Housz, already aware of the notion of oxygen, took up Priestley's observation and discovered that plants did indeed produce oxygen during the day, but that at night they produced carbon dioxide. He therefore introduced the notion of day-time and night-time respiration and, on the basis of his discovery, inferred that oxygen given off in the day-time arose from a decomposition of water. An astute observer, Ingen Housz also noted that the parts of the plant which were not green always produced carbon dioxide; he was then on the track of the role of chlorophyll (the pigment that makes plants green), but he did not follow up this excellent observation.

Green blood

One of the most unexpected turns in general biology was taken when the similarity was established, in 1894 by the Germans Schlunk and Marchlevski, between the haemoglobin of blood (which transports oxygen) and chlorophyll, both of which are derivatives of an aromatic compound pyrrole. This similarity, not without foundation and of particular interest in the history of the evolution of life, has only been mentioned incidentally by biologists.

The thread of these discoveries was taken up by the Swiss Senebier, who himself discovered in 1782 that a plant cannot live in an environment deprived of carbon dioxide.

In 1804, another Swiss, Saussure, established a connection between water absorbed by the sun, carbon dioxide and the growth of plants. In this he was repeating, probably without knowing, the empirical observation made by the Englishman Hales in 1724. (Hales had even had the remarkable intuition that plants took 'something' from the air.)

However, we were still a long way from understanding the fundamental concept of photosynthesis, a term which, moreover, would not be coined until 1898 by the Englishman, Barnes. A step towards that understanding was made, however, with another discovery, that of the Frenchman Garreaux in 1850: all living beings permanently practise the nocturnal type of respiration – that is, giving out carbon dioxide – but plants are the only forms of life which link respiration, or exchange of gases, and assimilation of those gases through chlorophyll.

The next major stage, the discovery of starch grains in chlorophyll, was made by the German von Mohr and began in 1845. From it stemmed an important observation: chlorophyll exists before the formation of the starch. It must be said that improvements in microscopes and the refinement of experimental procedures contributed greatly as time went on to the making of more discoveries. In 1857 the Frenchman Gris, having grown plants in conditions without light, discovered that the starch grains had disappeared.

Sachs' formula

These finds were not brought together until the German von Sachs made two important discoveries. First, chlorophyll is not diffuse in plants but is contained in small organelles (structures with cells) which would later be called chloroplasts; the plant does not synthesize starch where there are no chloroplasts. Second, starch is the first visible product of photosynthesis, but doubtless not the only one. Sachs also discovered that plants without chlorophyll do not synthesize starch, but other carbohydrates.

His studies lasted until 1864. Finally he submitted the chemical formula of photosynthesis:

$$6\,CO_2 + 6\,H_2O + \text{solar energy} \rightarrow C_6H_{12}O_6 + 6\,O_2$$

which partially explains the formation of carbohydrates such as starch. The same year, the Frenchman Boussingault completed the discovery by showing that the volume of oxygen released is equal to that of the carbon dioxide absorbed. Researches into photosynthesis, which could be said to be still incomplete in the 20th century, are progressing slowly towards refining our understanding of the singular mechanism through which plants ultimately use light to produce carbohydrates.

The last stages

The most outstanding stages were the discoveries of the phases of photosynthesis traced by the Russian Timiriasev (1877), the German Engelmann (1881), and then the Americans Blackmann (1905), Emerson (1921) and Arnold (1924). These botanists, among several others, established that chlorophyll first picks up solar energy, thus passing into a state of molecular agitation; one of the electrons escapes from the molecule and takes part in the hydrolysis of water, which starts a chain of chemical reactions preparing the second phase, in the course of which a sugar in the presence of carbon dioxide hydrolyzes to form other sugars.

In 1966 the American Calvin discovered another type of photosynthesis, called type C, maintained by two types of chlorophyll cells which allow some tropical plants to use the carbon dioxide in the air completely, and with less water than the usual type of photosynthesis.

Sun and sea

Studies in photosynthesis have been of obvious benefit to agronomy, but also to research into the use of solar energy, for example in the desalination of sea water. In two centuries, the journey so far has brought us a long way from Priestley's researches into 'good air'.

Pierced stone

Franco-Brazilian archaeological mission, 1973

Cro-Magnon in the Americas

The study of this Brazilian site
overturned old ideas about the
peoples of the Americas.

Until around the end of the 1970s, practically the whole world of international
archaeology maintained that the peopling
of the Americas occurred some 12 000 years
BC; it was probably brought about by
Northern Asiatic races, called Mongoloids,
who would have crossed the Iéna basin in
Siberia, then a land bridge now submerged
beneath the Bering Strait. A great many
anthropological studies confirmed beyond
doubt the Asiatic origin of the first American populations who probably colonized
the Americas in three successive waves, the
last being that of the ancestors of the present-day Eskimos.

A challenged thesis

Nevertheless, this 'classic' thesis did not
meet with the agreement of all archae-
logists, for one fundamental reason and

several other related ones. The first was that
the human remains whose dating is certain
go back 12 000 years, from Alaska to Tierra
del Fuego, and this is significant, for it
would mean that in a few centuries the
Mongoloids occupied the whole of the
Americas. This is doubtful since there are
signs that the immigrants observed pauses
in their migration of at least a few centuries.
The other reasons are that during the 1970s
remains which seemed to be more ancient
had been found. However, the 'orthodox'
theories were still maintained.

200 rock paintings

In 1973 a Franco-Brazilian archeological
team discovered a site, pointed out by the
Indians, at Boqueirado do Sitio da Pedra
Furada, or 'shelter of the valley of the
pierced stone', in the Brazilian state of
Piaui, deep in the jungle of the Mato
Grosso.

The site was exceptionally rich archae-
ologically, containing, amongst other
things, 200 examples of rock paintings,
more than were known in the whole of
Europe. The first datings carried out did
not contradict the classical theories as the
remains did not go back more than 12 000
years. Systematically excavated, however,
the site revealed several sedimentary layers
with a total depth of 3 metres. The lower
layers suggested an age greater than 12 000
years and the most ancient layer, situated
on a bed of rock, turned out, after carbon
14 dating (see p. 34) at the Centre des

Sailing boats in the age of stone?

In several prehistoric sites discovered in South
America wall paintings, called rupestrian, have
been found. These show mammals, fish, human
figures (of which certain seem to have their
arms held while they are beaten), women in
the process of giving birth, and what is even
more surprising, representations which seem
to be of sailing boats. As the existence of
navigation by sail in the Stone Age had been
completely unknown until now, specialists
hesitate to identify these pictures definitely as
boats.

Faibles Radioactivités du CNRS at Gif-sur-Yvette in 1985, to be 32 160 years old, give or take 100 years or so.

Considerable uproar followed, for this indicated, contrary to all the current ideas, that the first Americans were contemporaries of Cro-Magnon Man (see p. 49). Serious consideration therefore now had to be given to earlier indications of an occupation of the Americas much older than had been believed, such as the site of Tlapacoya in the Mexican basin, 24 000 years old and discovered in 1966; the El Cedral site, again in Mexico, 32 000 years old and discovered in 1981; and the Monte Verde site, 35 000 years old. At the time, however, the validity of the datings was challenged. This time the dating of the Pierced Stone site was carried out with elaborate precautions and submitted to numerous checks by the Frenchwomen Niède Guidon and Georgette Delibrias, and left no room for doubt.

In 1986 the archaeological world was still awaiting the publication of datings being carried out on remains found in South America and indicating a human presence 42 000 years ago.

In any case it appears that the crossing of the Bering Strait on foot could only have been possible during three periods of permanent glaciation: 62 000 years ago during an interval of 2 000 years; then 52 000 years ago for roughly the same length of time; and again 28 000 years ago for an interval of considerably greater length which ended some 13 000 years ago.

There is no sign that the crossing was made during the first of these three ice ages, but it certainly cannot be totally excluded. That leaves the second ice age, which obviously poses a problem, since the ice receded 10 centuries before the peopling of America.

Cro-Magnon or Neanderthal

Whatever the case may be, the importance here and now of these discoveries is considerable, for it has been shown that the Americas were occupied, not only in the later Palaeolithic period (10 000–35 000 years ago) but also in the middle Palaeolithic period (35 000–80 000 years ago). Since it is established that Cro-Magnon Man and his predecessor Neanderthal Man co-existed for some 20 000 years it might be wondered whether the same did not also occur in the Americas at the end of the middle Palaeolithic period. In this case, it would have to be supposed that there had been a first wave of Asiatic migrations of Neanderthal Men, followed by two waves of Cro-Magnon Men. It might equally be wondered whether Cro-Magnon Man, who up until now had been thought to have appeared at the beginning of the later Palaeolithic period, at least in Europe, might not have appeared earlier in Asia — 40 000 or 45 000 years ago for instance — at the end of the middle Paleolithic period. That being the case, there would be no need to assume a first migration of Neanderthal Men and many theories about prehistory would have to be revised.

An essential discovery in palaeontology

Whatever the case, the discovery of the Pierced Stone stands as one of the most important in palaeontology. Doubtless it represents only a beginning, not only in the history of the populating of the Americas, but also in the origins of the human race. We can expect, in fact, that the American territories, which have so far been very little exploited by archaeologists, will be the objects of numerous explorations and will reveal yet more new sites.

Pluto

Tombaugh, 1930

The last of the planets

It is not yet known whether this is a planet or indeed a satellite which has escaped from Neptune's attraction.

The discovery of the planet Pluto is one of the strangest in the history of the sciences, since it appeared at first to be deliberate, while in fact it was merely accidental.

At the beginning of the century the American astronomer Lowell noted that the orbits of Neptune and Uranus demonstrated irregularities, which he explained by the gravitational influence of an unknown planet lying beyond their orbits which he named X. His reasoning was similar to that of Leverrier and Adams, who mathematically deduced the existence of Neptune and Uranus before their effective discovery (see p. 208). Moreover, the calculations of other astronomers agreed with Lowell's.

He instigated a series of observations. On his death in 1915 the research into Planet X was entrusted to his successor, Clyde Tombaugh. A telescope was constructed specially for that purpose at the Lowell observatory in Arizona; it was only brought into service in 1929. In January and February 1930 Tombaugh located, on several photographic plates, a celestial body which was indeed a planet and whose movement corresponded closely with Lowell's predictions. The announcement of the discovery of Planet X, later called Pluto, was made on 13 March 1930, on the anniversary of Lowell's death. However, the American Brown showed that Lowell's forecasts were not valid from the dynamical point of view.

It appeared that Tombaugh had indeed discovered a planet by taking false calculations as his basis. The debate ended with a further point made by the American Bower, who calculated that on the one hand the mass of Pluto was half that of Earth and that, on the other, the orbital anomalies of Neptune and Uranus had been overestimated. With this new mass of Pluto the corrected anomalies of the two other planets could not be explained. This is a further example of false or half-right theories leading to a real discovery.

Pluto remains the most mysterious planet of the solar system. Its density seems very low (perhaps only 1 per cent that of Earth) and does not correspond absolutely to the dynamic calculations of its mass which, contrary to Bower's claims, would represent less than one-fifth of that of Earth. Its strongly eccentric orbit and the very accentuated tilt of its rotational axis give rise to the idea that this was a satellite of Neptune which escaped from its primary orbit. Indeed, Pluto's orbit lies inside that of Neptune for a small way. Information about Pluto was transmitted by the space ship Voyager II at the time of its passing close by the planet in January 1987.

> **Scepticism**
> The interest being shown in this discovery is considerable, but is clouded by the scepticism of several astronomers. Pluto's brilliance being ten times less than had been supposed, these critics infer that its mass is proportionally less and that a body that small could not be sufficient explanation for the orbital anomalies of Neptune and Uranus.

Polonium and radium

Pierre and Marie Curie, 1898

'Explosive' rocks

The discovery of these radioactive elements was received with scepticism.

At the time of her research into radioactivity (see p. 133), when she analyzed one by one the rocks drawn from the mineral collection of the Ecole de Physique, rue Lhomond, in Paris, Marie Curie found examples of pitchblende which produced much more radioactivity than expected, given their percentage of uranium and thorium. At first she thought there must have been an error in her measurements and repeated some of them up to 20 times, but in the end the weight of evidence prevailed: there existed in these samples substances much more radioactive than uranium and thorium. These were unknown substances, for after having checked out all the listed elements, she found none which released as much radioactivity.

A mystery

She raised this mystery with physicists who advised her to be cautious, as the measurements might yet be wrong, even though they had been repeated many times. After several months, however, the evidence could not be denied and on 12 April 1898 Professor Lippman presented a communication from Marie Sklodowska-Curie about her observations to the Academy of Sciences. In it the young woman wrote: 'Two ores from uranium, pitchblende (uranium oxide) and chalcolyte (phosphate of copper and uranyl), are much more active than uranium itself. This fact is very remarkable ...'

Pierre Curie collaborated in his wife's research. The two scholars proceeded to analyse chemically all the elements of pitchblende in order to isolate the mysterious radioactive element or elements. By a process of elimination they arrived at two unknown parts. In July 1898 they were sure that they had isolated at least one element. In honour of her country of origin Marie Curie called it polonium. It took the 84th place in Mendeleyev's periodic table of elements (an empirical way of ordering the known elements by their molecular weight).

The basis of radioactive chemistry

The method of analysis thought out by Marie Curie for her research is still the basis for radioactive chemistry today: consists of measuring the specific radioactivity of each

---

A prophet in his own country ...

When, in 1902, the Curies succeeded in isolating radium in its pure state, they acquired international renown. The greatest scientists of the time, Crookes, Boltzmann and Rutherford, wrote to them. In France their working circumstances remained precarious and, when Pierre Curie asked for a larger laboratory, it was refused. When he applied for election to the Academy of Sciences the unknown Amagat was preferred to him. He was offered the *Légion d'honneur*; he refused it, and again asked for larger premises which were refused him once more. In 1903 the Curies and Becquerel were nominated for the Nobel prize for physics. In 1904 the Chamber of Deputies alloted the Curies one-fifth of the sum needed for the setting up of their premises. Pierre Curie was to die in 1906 without ever having his laboratory.

of the constituents of an ore. For this purpose she designed a new piece of equipment where the ionizing charges were collected on a tray in an ionization chamber and balanced by mechanical charges developed on a lamina of quartz (piezoelectric charges). An electrometer then enabled the ionization to be measured. The equipment had been conceived by Jacques and Pierre Curie, thanks to a new phenomenon they had discovered: this was that quartz subjected to mechanical pressure produced electricity. On 20 December 1898 the Curies isolated another element, radium, which occupied the 88th place in Mendeleyev's table.

This first-rate research was carried out in the same deplorable setting in which the Curies discovered radioactivity. This in itself had so contaminated the environment and measuring equipment that the work became increasingly difficult, not to mention the damage it was doing to the health of the researchers.

The Academy of Sciences published the Curies' discoveries but the scientific world remained sceptical. No-one had yet in fact seen polonium or radium in their pure form. The Curies were therefore to turn their attention to isolating these substances. This was difficult, for pitchblende, the substance in which they hoped to isolate the two elements, was a relatively expensive ore which was only produced from one known mine, at Joachimsthal in Austria; it was there that a business had been set up to extract the uranium oxide salts used at the time in the glass-making industry.

The Curies imagined that if only the salts of uranium oxide were extracted, the residue ought still to contain the radium and polonium they were interested in, and that this residue, being of no industrial use, ought not to be very expensive. So the Curies planned to buy, from their meagre budget, a few hundred kilograms of this residue, for, as Eve Curie was later to write, if they had asked for official funding the authorities would have laughed in their faces. Interventions on their behalf by sources close to the Austrian government resulted in its graciously offering a ton of pitchblende to the two French scholars.

For handling the 1000 kg of brownish rocks which had been sent to them in linen bags the Curies were to find their circumstances no better looked on by the authorities than they were when they were first looking for a place to undertake their work on radioactivity. They were allocated a wooden shed. It was there that from 1898 to 1902, dressed as labourers, in an atmosphere of toxic dust which covered their clothes, skin and hair, the Curies explored one of the greatest discoveries of physics. Their only help in transporting the bags of pitchblende was a laboratory assistant by the name of Petit. In 1900 Pierre Curie made the acquaintance of a young chemist, André Debierne, who was also studying radioactivity (in 1899 he had isolated actinium) and who was to liaise with the Curies in their work.

The dawn of the atomic era

Nevertheless it was with only her husband's help that Marie Curie worked through the pitchblende kilogram by kilogram, until the moment when the two scholars arrived at the end point: purification and fractional crystallization. At the beginning they had anticipated that they would find 1 per cent of radium per ton of pitchblende but they realized that they would in fact obtain much less than that. In 1902 the exhausting work of 45 months had yielded one-tenth of a gram of pure radium. This news astounded the scientific world, which had no way of knowing that the fragment with the bluish sheen which glowed in a box in a miserable shack in the rue Lhomond heralded the atomic age.

However, several scholars had grasped the importance of what Marie Curie had said to the Academy of Sciences in 1898 and had, like Debierne, thrown themselves

> **The first woman to teach at the Sorbonne**
> On the death of Pierre Curie in a traffic accident on 19 April 1906, Marie Curie succeeded him in the chair which he held at the Sorbonne; she therefore became, on 13 May that same year, the first woman to teach at the Sorbonne. Her position was made permanent two years later and she continued her research.

into research. So it was that in the same year, 1902, the Germans Elster, Geitel, Hoffmann and Strauss discovered radium D, an isotope then called 'radio-lead'. In 1905 another German, Otto Hahn, discovered radiothorium and mesothorium; in 1907 the Englishman Boltwood discovered ionium. Chemistry of the radio-elements (as the newly found radioactive elements were known) progressed rapidly for, in 1905, the Englishman Thomson discovered weakly radioactive atoms in potassium and, the following year, his compatriots Campbell and Wood discovered others in rubidium.

Rapid progress

The discovery of radium in particular opened up immense perspectives for scientists. At the beginning of 1903, Curie and Laborde stated that radium spontaneously gave off heat, which, they noted, 'cannot be explained by an ordinary chemical change ... but by a modification of the radium atom itself. A similar change, if it exists, occurs extremely slowly, for the properties of radium experience no notable variation in several years. So if the preceding hypothesis were correct the energy brought into play in the transformation of the atoms would be extraordinarily great.'

The hypothesis was correct, as the turn of events was to show. It was certainly reinforced, if only intuitively in Curie and Laborde, by the verifications of Becquerel and then of Marie Curie. The former had already cast doubt on the indivisibility of the atom, on the basis of photographic plates dimmed by clusters of particles emitted by uranium. Atoms which permanently lose so many particles cannot in themselves be indivisible. Marie Curie did the same experiment with radium.

While his wife persisted in the work of extraction of the radio-elements, Pierre Curie turned to the study of their physical and luminous characteristics. It was in this way, while measuring the action of magnetic fields on rays emitted by radium, that he discovered that there were some which were electrically positive, others negative and still others neutral. Taking up this particular discovery, the Englishman Rutherford verified it by using his previous work on X-rays. It was he who had distinguished alpha rays, powerful but not very penetrating, and beta rays, less powerful but much more penetrating.

Not until 1932 would the Englishman Walton prove the possibility of fission of matter by an experiment bombarding lithium with a bundle of hydrogen protons: the lithium captures a proton, and its nucleus, going from a mass of 7 to a mass of 8, divides into two alpha particles of a mass of 4, while liberating 8 million electron-volts of energy. The world was now only a few years away from the first A-bomb.

The only woman to have received two Nobel prizes

In 1910, the year that her treatise on radioactivity appeared, Marie Curie succeeded for the first time in isolating metallic radium in a pure state, a feat which was mentioned in her list of achievements when the Nobel prize for chemistry was awarded to her in 1911. This was her second Nobel prize, since she had shared with her husband and Henri Becquerel the Nobel prize for physics in 1903 for the discovery of radioactivity. She is therefore the only woman who has twice been honoured by the prestigious Nobel jury.

Polymerization

Staudinger, 1922; Carothers, 1930

The invention of a discovery

*The combining of small molecules
into big molecules has been the
starting point for all synthetic
materials.*

Polymerization — that is, the combining of small molecules (or monomers) into large molecules (or polymers) — stands halfway between invention and discovery. It was an invention, because the theory was first put forward in 1922 by the German Staudinger, the first to postulate that small molecules can combine together to form large molecules, not by simple physical alignment but linked by atomic bonds to form chains or compounds. It was in his communication of May 1922 that he first used the term polymer, in an analysis of rubber (see p. 189). Staudinger would show that this substance is formed from a chain of combined molecules of isoprene. The theory began therefore with a discovery, but it was strongly contested, which led Staudinger to take up his research again. This led him to a second discovery, made this time concerning the resins of styrene: these consisted of macromolecules or giant polymers of different lengths, which he called polystyrene. The theory was gradually accepted.

The first plastics

Synthetic materials had actually existed for three-quarters of a century, but they were not particularly satisfactory (see p. 199). Parkesine, which became xylonite, was extremely flammable and, when it became celluloid after modification of the manufacturing process (by the addition of camphor), it still remained flammable. The first entirely synthetic plastic, Bakelite (so called after its creator Baekeland, an American of Belgian origin) had been discovered in 1910 and patented in 1918. However, its chemical structure was unknown; it was certainly a compound of urea and formaldehyde polymerized by heat, but that is as far as our knowledge extended at the time. Produced commercially several years after the formation of Staudinger's theory, it too proved to be less than entirely satisfactory.

Ersatz products

Certainly these were not the only plastics on the market, since vinyl chloride had been in existence for almost a century, and polyvinyl chloride had been created in 1912. The polymethyl methacrylates had also been polymerized experimentally since 1880, obviously without it being known what was happening when they were heated.

In the end, we had to wait for Staudinger to reveal what actually occurred when plastics were manufactured. Doubtless the achievements made by this scholar would have been disregarded were it not for the fact that Germany was actively pursuing its own industrial reconstruction and that the manufacture of synthetic materials, the famous *Ersatz* products, enabled Germany to enhance its economic independence from the producers under the control of the great colonial powers. Staudinger's lesson was quickly understood and one of his discoveries in particular caught the attention of another researcher, the American Carothers: this was the existence of a connection between the molecular weight of a polymer and its viscosity.

Nylon

While German industry was perfecting its production of polystyrene, American firms were also continuing their research. Carothers, named chief of research at Du Pont De Nemours & Co. in 1928, was studying the structures of long-chain molecules when chance took a hand. Having left a preparation heating for too long (it was essentially a base of carbon, hydrogen, nitrogen and oxygen), he discovered that it was producing fibres capable of being drawn out to a great length before breaking. It would be going far too far to pretend that it was entirely by chance that polyamide, later called nylon, was discovered. Carothers was exploring the whole area as a specialist of the first rank and chance merely gave him a slight nudge in the right direction.

A biological phenomenon

The modification of formulae and treatments expanded polymerization until it became recognized as an autonomous field in organic chemistry. Whatever the variety of combinations and techniques, one principle remained unchanged: that of the functional nature of molecules. Each molecule of a polymer must comprise at least two reactive groups, such as an acid (-COOH) or an amino (-NH_2) ending. This is so the individual molecules can combine with each other to form long chains, linking together via their reactive groups.

Although known for its role in the manufacture of synthetic materials, polymerization is by no means limited to that area; it is a phenomenon which plays an essential role in biology. The construction of long-chain proteins from amino acids is also characterized by polymerization. The formation of the first living molecules, probably unicellular, is explained by a polymerization of amino acids which occurs in particular geophysical conditions, involving the presence of absorbant minerals, a specific temperature and the diffusion of ultraviolet rays. (The greater part of organic polymerizations in fact involves an initial dehydration: that is, the loss of one molecule of water at the time of the joining of two monomers.) Furthermore, in 1965 the American Sydney Fox completed a masterly experiment on polymerization of amino acids, to demonstrate that the formation of the first proteins was due to chance.

The greater knowledge of polymerization acquired since the discoveries of Staudinger and Carothers has also greatly encouraged theories of the possibility of life, albeit primitive, on other planets, through the formation of large organic living or 'pre-living' molecules. Some theoreticians even propose sowing planets with macromolecules in order to start off a process of life formation inspired by that which prevailed on Earth.

From hair to diamonds

Even the term 'polymers' immediately evokes in the layman the idea of synthetic substances. However, it should be remembered that it describes complex molecules made up of units of low molecular weight, and that it also indicates natural organic or mineral substances, as well as artificial ones. The proteins and nucleic acids, nails and hair are polymers, just as diamonds, sulphur and selenium are.

Synthetic polymers

Synthetic polymers comprise seven groups: the polyhydrocarbonates like polythylene; the polyvinyls and polyacrylates; the polyacetates and polyethers; the polyesters; the polyamides; the polyurethanes; and the mixed organic-inorganic polymers like the silicones.

Prion

Prusiner, 1982

A virus without a 'soul'

*The existence of organisms without
DNA represents one of biology's
greatest challenges.*

In 1982 the American Stanley Prusiner
made a discovery of the highest importance
in the field of biology: infectious agents with
no genetic material, therefore apparently
incapable of reproducing and yet reproducing themselves. They were simple proteins. He called them 'proteinaceous
infectious particles', from which the name
prion is derived, as a kind of acronym.

Kuru

Prusiner's discovery, with which Hadlow
and Eklund were also associated, sprang
from work on an illness which had been
known for nearly two centuries, a nervous
disease in sheep called scrapie. It had been
known to be an infectious disease since the
Frenchmen Chelle and Guille had shown
in 1935 that a healthy animal could be
infected with it by being injected with
material from the brain of a sick animal.
The agent of the disease had not been
isolated, however. In 1959, Hadlow, later
to collaborate with Prusiner, had linked the
illness to an infection endemic in Papua
New Guinea, called Kuru. In fact, two
years before, Zigas and Gajdusek had established that the illness was rife only among
the cannibals of that island, who would eat
the brains of a dead parent as an act of
respect. The disease was being transmitted
because the dead were continually being
consumed by the living. When this practice
ceased, kuru disappeared.

A slow virus disease (its evolution can last
twenty years), Kuru attacks the brain just

as scrapie does in sheep, with the difference
that the latter only seems to be active
among animals and the former among
humans. There are striking points of similarity. The difference in the spheres of infection disappeared when it was shown that
kuru could also infect monkeys, just like
another disease whose origin is a mystery:
Creuzfeldt–Jacob disease, which also
attacks the brain and like kuru causes early
dementia. The three diseases therefore had
one common point, especially as in the three
diseases infection did not produce inflamm-
ation and was not characterized by a
deterioration of the cerebrospinal fluid
which surrounds the brain and spinal cord.
This indicated, according to Prusiner's
findings, that it was not producing any reaction in the immune system.

The discovery, in hamsters, of a form of
scrapie with a rapid incubation period and
the invention of new techniques of analysis

A curious disagreement

In 1978 the American Pat Merz had
photographed, in sections of brain from mice
affected by scrapie, strange fibrilla which she
supposed to be agents of the disease. These
fibrilla resemble the prions photographed by
Prusiner in the brains of hamsters affected by
cerebral degeneration, but Prusiner
energetically denied that the two types of fibrilla
were one and the same. That, however, was
not the opinion of the scientific community in
general, which is, moreover, extremely divided
on the question of prion.

allowed for the isolation and investigation of a type of protein called a glycoprotein, which is half the size of a red blood cell and has a tendency to join together in rods.

The puzzle was taking shape, for a protein cannot reproduce itself without genetic coding, yet neither DNA nor RNA were found in it (see p. 52). Besides, a particle possessing genetic material is, almost of necessity, sensitive to ionizing radiation. When, in 1964, the British scientists Alper, Haig and Clarke had irradiated tissues infected with scrapie, not yet isolated obviously, they noted that this treatment did not render the tissues less infectious, thereby concluding that there was no genetic material present. Yet the idea of an infectious agent bereft of genetic material was unacceptable, and their work and conclusions were greeted with scepticism. When Prusiner and his collaborators irradiated infected brain tissue in their turn, however, they obtained the same results.

Prion is composed of some 250 amino acids. To suppose that these could gather together and take shape without genetic indicators, that is without coding by *nucleotides*, (see p. 53) would be to regress to admitting the possibility of spontaneous generation, which is absurd.

Three explanations

Prusiner therefore offered three possible explanations. The first was that prion did possess, not a DNA but a RNA messenger; this makes prion an example of a retrovirus. This hypothesis hardly tallies with the data gathered from observations of prion. The second possibility was that it might be the proteins themselves which, in some way, fulfil the role of RNA, that is they are matrices for the production of new proteins useful for replication. That would be a method of production unknown in biology and, furthermore, very difficult to imagine. The third explanation, favoured by Prusiner, was that prion activates or changes a gene of DNA from the host cell, in such a way that this codes the cell for production of the protein. All that would be required would be for prion to contain an element of nucleic acid, small enough for it to have escaped observation until now.

In this case prion would be behaving in a similar way to viruses, but with considerably reduced genetic material. (Viruses reproduce themselves by inserting some of their genetic material into that of the host cell.) In any case the existence of several types of prion is difficult to reconcile with the activation of a specific gene. What we were left with was the following hypothesis: that the simple protein of prion might bond with DNA from the host cell in a way which would activate the transcription of one or several genes.

French research

French research has, however, modified certain of Prusiner's hypotheses. Starting from the work of the American Merz, who discovered microscopic threadlike structures or fibrilla infesting the nervous system of patients suffering from Creuzfeldt–Jacob disease, Latarjet, of the Curie Institute, indicated that these fibrilla might in fact be chains of DNA filaments surrounded by a protein casing. In a separate development, fragments of extra RNA have been discovered in samples of infected organs; these fragments might reveal an infectious origin. In that case, prion would be a particular type of retrovirus having a very fine genetic material, which would explain how it had escaped microscopic detection, at least until 1986. If that were the case, the role of the

Enzymatic RNA and prion

In 1981 the American T. Cech made a discovery which he was to take further in 1985 with his colleague A. Zong: this was that RNA, a nucleic acid which had until then been thought to be closely bound up with DNA for the synthesis of proteins, was in fact an enzyme and could in certain cases synthesize proteins alone. This was extraordinary discovery, which may clear up the mystery of prion. No DNA has in fact been found which explains its synthesis, but that does not exclude the possibility of its containing fragments of enzymatic RNA which would be sufficient to ensure its replication. If the hypothesis, envisaged by certain biologists, was verified, it would lead to a considerable rethink of key ideas in biology.

proteins discovered by Prusiner and identified as being prions would still have to be determined. This would call for a new theory: namely that this hypothetical retrovirus produces proteins, a phenomenon of which there is no other known example.

A pseudo-virus?

The discovery of prion represents a decisive event in the history of slow viruses in general (a family to which the HIV or human immunodeficiency virus of AIDS belongs) and of Creuzfeldt–Jacob disease in particular (as well as a related disease, Gertmann–Strassler syndrome). It has allowed us to establish that there are infectious diseases which do not trigger an immune reaction. In the case of AIDS it is known that the absence of this reaction is due to the fact that the virus infects exactly those white blood cells (lymphocytes) which bear the immune defences, the T4s. In the case of prion infections we are probably concerned with another mechanism, since both Prusiner and the Koch Institute in Berlin succeeded in producing experimentally antibodies against the infectious agent.

It will therefore be of prime importance to establish the way in which infection by the prion type of virus can be established, and if this virus is the only one of its kind. Finally, it will be equally important to establish just how viruses, formerly thought of as affecting only animals, can also affect humans.

Prusiner's discovery was made, it should be noted, at almost the same time as the American Gallo and the Frenchman Montagnier were establishing that the AIDS virus is a retrovirus, therefore another member of a family which had long been supposed to infect only animals.

In the same way, Prusiner's discovery must be related to that of the Americans Sodroski, Wei Chun Goh, Rosen, Campbell and Haseltine in 1986, which seemed to indicate that proteins alone can have infectious effects. Trying to produce a vaccine against AIDS, using the outer envelope of the virus responsible, HIV, this team discovered that the covering, still composed entirely of proteins, affected human lymphocytes in such as way as to make them inoperative; this is odd, given that in principle the viral covering has no genetic material and cannot therefore infect cells.

Finally, even if the incidence of Creuzfeldt–Jacob disease is relatively low, the identification of its agent gives rise to the hope that a cure will be available in the fairly near future. As has already been seen, it is possible to produce antibodies against prion; at the end of 1986 work with a therapeutic aim was being pursued just as actively as the purely virological research in the field of prion infections.

Prion, virion or SAF?

The name 'prion' has not been universally adopted by specialists, given that it implies that the responsible agent would be exclusively a protein, while there are indications that there might be elements of RNA. The name 'virion' has been suggested as an alternative, but it seems more likely that the term 'SAF', for scrapie-associated fibrils will eventually take over from the term 'prion'.

Retrovirus

A virus with an outer envelope enclosing the core, its genetic information is stored in a molecule of single-stranded risonucleic acid (RNA). Retroviruses include the *oncogenic* (tumour-forming) viruses.

Priore's 'machine'

Priore, 1948–83

The Priore affair

An incomplete discovery of the 20th century: a machine which cures cancer by electromagnetic radiation.

One of the most intriguing discoveries in the history of the sciences is the one which, between 1948 and about 1960, culminated in the invention of a piece of equipment capable of curing certain cancers and diseases, if not cancer in general.

The discovery is intriguing because it was achieved by a self-taught man, of which there have been many, at least in the sciences of the last century; this one was a distrustful person who never revealed the secret of his equipment. Unfortunately it did not lead to a development which could be of great value to public health; instead it sank into a succession of mishaps and often obscure intrigues which, after the death of the inventor in 1983, totally blocked all research and seriously compromised some great scientific reputations.

Curing cancer

In 1948 an Italian electrical engineer, Antoine Priore, having 'established' that an orange placed in an electromagnetic field did not go rotten, decided to set up an electromagnetic field generator at Floirac, near Bordeaux, in order to study the possible effects on living tissues. The equipment was cobbled together, Priore acting as an amateur and freelancer, without any official backing; it was not even a question of an invention in the classic sense of the word, since electromagnetic field generators already existed and since in the circumstances Priore was not exploiting particular knowledge for anticipated ends. He groped around and set out on the discovery of the unknown, armed with knowledge gleaned from a work dating from 1928, *Cancer, electrical intervention*, by Charles Laville, and from chance readings. The extraordinary aspect of what was to become the 'Priore affair' was that this electrical engineer took it into his head to cure cancer.

How many cures?

After some initial experiments in animals, Priore decided to progress to humans, and did in fact treat them, sometimes secretly, and sometimes with the agreement of doctors who sent desperate cases to him. Files do not exist for all these cases, but a few are available for study. One refers to a boy of 12 suffering from Hodgkin's sarcoma, a serious form of cancer, treated in 1954 and who appeared, in 1966, to have been cured. Another case is of a man suffering from the onset of cancer of the larynx in 1955 who, after treatment with 'Priore's machine', saw the cancer recede after two months. It seems that Priore treated dozens of cases at least, perhaps more, but his files are not accessible. It is therefore impossible to know what percentage of cures he may have achieved.

It is often difficult to make a scientific judgment on the nature of a cure in cases treated by electromagnetic radiation. Such cases, like those of miraculous cures at Lourdes, need to be followed up over several years according to the usual procedures of oncology (study of tumours).

An official experiment

At the beginning of the 1960s, Priore (who was still treating patients without charge) realized that his 'machine' would never fulfil its great potential without the endorsement of the scientific establishment. After a few snubs, the scientific world, in the person of the Vice-Dean of the Faculty of Medicine at Bordeaux, gave its agreement to a scientific experiment, on animals of course (although the treatment of humans was continuing in secret). The designated animals were to be rats, on to which had been grafted a type of incurable tumour called T8. It was revealed that, in the rats treated by the machine, the mean volume of the tumours was 60 per cent less than in those rats given the classical X-ray treatment, and that these tumours only rarely spread to other parts of the body.

The first setbacks

In spite of the growing hostility of certain doctors of renown, due as much to the mysterious character of the 'machine' as to the risky nature of the effects of electromagnetic radiation on living tissues, an area ripe for speculation, Jacques Chaban-Delmas, Mayor of Bordeaux, set up, in 1961, two successive commissions to analyse Priore's discovery. The two commissions rejected the documents. The setbacks had begun and with them, the intrigues.

Two researchers, Biraben and Delmon of the Bordeaux faculty, cured rats who were carrying the previously mentioned T8 tumour but could not contravene an injunction which had been served on them (in a manifestly unscientific spirit) forbidding publication. News of their success spread; two other researchers, this time from the Gustave-Roussy Institute at Villejuif, repeated the experiment with the machine and arrived at the same results: the incurable tumour T8 was indeed cured, as was (still in animals) another equally incurable disease, lymphoblastic lymphosarcoma 347. All that is known of the technical conditions in which the experiments were carried out is that the magnetic fields used had strengths of 300 and 620 gauss.

Division in the Academy of Sciences

The rest of the affair reads like a novel. Eminent personalities in medicine, such as Professor Robert Courrier, Permanent Secretary of the Academy of Sciences, Professor Raymond Pautrizel, Director of the Immunology and Parasitic Biology Laboratory of the medical faculty in Paris, Professor Antoine Lacassagne and many others, manned the battlements; but a veritable band of guerrillas divided the Academy of Sciences and the scientific world as a whole. The adversaries had an important epistemological argument: the supporters of Priore published results achieved with a machine whose nature they did not know. Their methodology was risky, a claim which could be defended, but only by ignoring an elementary fact: even if Priore's machine was unknown, that did not prevent explicit results from being obtained.

A lively polemic

A cabal was formed, supported by the then prestigious organs of the press. The scientists, like Professor Pautrizel, who continued

---

Proof of magnetotaxis in bacteria

The first formal observation of the influence of magnetic fields on eukaryote micro-organisms (those possessing a cell nucleus) only occurred in 1981 with the Americans R.P. Blakemore and R.B. Frankel. It brought to light the fact that certain microscopic green algae follow the force lines of the Earth's magnetic field. In 1974 this same Blakemore had discovered a similar behaviour called magnetotaxis in prokaryote bacteria (those with no cell nucleus). The magnetotaxis of these lower organisms could partially explain magnetotaxis in vertebrates like pigeons, mice, salamanders and bees. It confirms the sensitivity of living organisms to magnetic fields in general.

The carbon chains of fatty acids are modified. The consumption of water subjected to the same fields leads to an increase in the unsaturated fatty acids in the adipose tissues of mice, guinea pigs and hens.

to show an interest in Priore's machine did so at the risk of their careers. It was one of the most deplorable internal quarrels in the scientific world since 1887, when the inept Professor Peter accused Pasteur before the Academy of Medicine of giving rabies to the people he vaccinated and the journalists Rochefort and Meunier repeated the unworthy accusation.

Difficult technical and financial dealings added to the weight of hostility against the machine, the fifth example of which (Priore had had four constructed), the M235, did not work. Tens of millions of pounds invested by governments and private firms were completely written off.

Strangely enough, Pautrizel did manage in 1966 to cure a parasitic disease, trypanosomiasis, by exposure to electromagnetic radiation. Nothing came of this; the hostility of the majority of scholars had reached a critical point and, when Priore died in 1983, the 'affair' died too. The machine was buried.

The exact facts

Three facts remain. The first is that Priore did make a discovery of great importance, as the results in animals and humans show.

The second is that his machine was a generator of electromagnetic waves and magnetic fields modulated by a discharge tube containing an 'inert gas', supplied with a special cathode and anode (negative and positive electrodes), activators with continuous electric fields and alternating fields of high and low frequencies, axial and transverse magnetic fields created from a variable frequency modulator, and a continuous source of current of medium

---

The body's natural electromagnetic fields

In 1967 the Russian Gulyaev, of the State University of Leningrad, discovered that the heart, nerves and muscles of the frog, and the heart and muscles of humans produce a weak electromagnetic field. This field is explained by the nerve-impulse which runs through these tissues. Following this, it was discovered that such fields existed around numerous other human organs, including the brain.

---

power. It seems that Priore experienced two difficulties: the first in the increase of the power of his electromagnetic field, and the second in the marking out of certain apparently more effective frequencies between 14 and 16 million Hertz (wavelengths of 19–21 m).

The third fact is, it would seem that 'Priore's machine' did not cure the cancer, but considerably reinforced the immunological defences of the organism. That is at least what is indicated by the repeated rejection of grafts by the mice under treatment, something which for a long time caused fraud to be suspected. (A substitution of animals was thought likely, for it is not normal for animals of a given family to reject grafts from their own family.)

One of the problems in scientific acceptance of Priore's machine was no doubt the jealous secrecy with which the inventor surrounded it. Inexperienced in the ways of science, Priore did not grasp that scientists are reluctant to take into consideration any experiment where they are not in control of all its elements. He merely planned to demonstrate the efficacy of his discovery and manufacture dozens of machines under the seal of secrecy. He was afraid that his secret might be stolen from him.

No scientific luggage

Priore's unpredictable character and his very mediocre scientific baggage brought about his downfall. A discovery cannot be defended by someone who neither understands nor masters its data. Not only was Priore ignorant of biology, not to mention the study of cancer — his notes on the subject are utter gibberish — but in addition he did not even master the data of his own discovery; witness the failure of the fifth 'machine' and the constant tentative fumbling which he was given to, even during scientific experiments, to the great irritation of those taking part.

Finally, we can be in agreement with certain of his adversaries on the following point: electromagnetic radiation can be dangerous, as all its users know. It can even kill.

If there is little doubt that Priore did accidentally discover certain effects of electromagnetic fields on cellular disorders, he was certainly not the only discoverer in this field; the particular interest of his discovery stems from its application in the treatment of human illnesses, notably cancer. In 1938 the Dutchman Van Everdingen had noted the inhibition of tumours in rats subjected to electromagnetic frequencies of 1870 and 3000 million Hertz (wavelengths of 10–16 cm), with return of the growth at the end of the treatment; and, in 1949, the Italian Montani claimed the complete resorption of sarcoma (a type of cancer) in a rat after exposure to frequencies of 600 million Hertz (50 cm) of very weak intensity, results confirmed by the Englishmen Roberts and Cook with frequencies of 3000 and 10 000 million Hertz (3–10 cm, i.e. microwaves).

The 'incurable' cancer, cured in 1961 by Guérin and Rivière of Villejuif with the help of a Priore machine, the same lymphoblastic lymphosarcoma 347, had already been treated with positive results in 1941 by van Everdingen.

Ill-fated effects

There is a considerable body of data on the effects of electromagnetic fields on living tissues, dating from both before and after Priore's discovery: from the modification of the electroencephalogram (see p. 59) to the modification of the blood formula, from the increase of glycaemia to the inhibition of enzyme activity, they are all negative. Magnetic fields almost always seem to induce pathological (harmful) effects, and electromagnetic fields disturb just about the whole biological equilibrium whenever they reach intensities and frequencies higher than the norm in the environment.

One of the reasons for the pathological effects could be their influence on the electromagnetic control mechanisms governed by the nervous system and by simple biochemical cellular changes. The other reason for the unfortunate effects, this time from electromagnetic fields only, is that they can considerably raise the temperatures of tissues (in the same way that microwave ovens cook food; see p. 126). Prolonged exposures of animals to such fields have in fact led to fatalities, but this may be exactly the effect which brought about regression in the cancers treated with Priore's machine. The treatment of cancer by local increase in temperature (hyperthermia) has, in France at least, become part of accepted hospital practice.

The machine in limbo

The whole affair of Priore and his machine must take its place in the history of those discoveries which are like gropings in the dark. Quite apart from the indisputable skills which were mobilized for the study of the machine's magnetic and electromagnetic effects, it is beyond doubt that this study was compromised from the start by serious errors, and by an erratic methodology, since none of the researchers knew specifically whether Priore used either a magnetic or electromagnetic field, or both, nor their frequency or intensity.

An apparatus like Priore's would not be impossible to construct, although it would be a slow and tricky task; but in any case this whole area is tricky, since one is looking for effects which are unknown. Possibly, in years to come, state or private funds will be found to look again, in a really scientific way, at the effects of magnetic and electromagnetic radiation. It is perhaps surprising that this has not been done already, for there is no shortage of work pertaining to these effects, quite independent of those relating to Priore's machine. We have already seen, though, that great discoveries can spend a varying length of time 'in limbo'.

Theoretical physics and organic mysteries

It is probable that if the method of transmission of electromagnetic waves were known, then their effect on living tissues would be better understood. The one effect that is known for certain is that they produce heat, just like a microwave oven. One could therefore reduce their anticancer action, in the case of Priore's machine, to a thermal action. However, specialists who have taken an interest in this machine think that heat alone is not in itself sufficient explanation for the results which have been recorded.

Protozoa

Leeuwenhoek, 1675

Leeuwenhoek's little zoo

It took two centuries before the nature and importance of these microscopic unicellular organisms was established.

The third great discovery, after bacteria and spermatozoa, (see pp. 102 and 193) which made the Dutchman Van Leeuwenhoek worthy of a place in posterity was that of protozoa. His carefully described observations of these unicellular organisms, almost all microscopic, created a lively interest at the time but then passed almost unnoticed for nearly two centuries in the biology and medical establishments. The former would have been able to acquire from it more precise notions about life; the latter would have been able to understand the origin of the parasitic diseases raging then as now (for example, amoebiaisis and malaria).

It was the Englishman Robert Brown who, in 1833, returned to his precursor's tradition of observation, revealing that the cellular nuclei which he had identified in plant cells, also existed in animal cells, including those of protozoa; he found that,

in the latter, they were not encased in cellulose, an observation which might today seem naive but which had the value of differentiating between the animal and vegetable kingdoms.

Cellular division

In 1839 the German biologists Theodore Schwann and Mathias Schleiden put forward the principle according to which cells are the rudimentary particles of the organism; they also observed that certain organisms are multicellular, others unicellular. As a matter of fact their theories were considered to be the synthesis of ideas generally accepted at the time.

As biology was only in its early stages, ideas on the origin of the cell were rather confused, nowadays for instance the theory of 'cytoblastism' postulated a hypothetical preformed substance out of which developed the nucleus which then gave birth to the cell. The microscopic observation of protozoa, had it been practised, would doubtless have helped to correct this metaphysical interpretation of life. In such a context the discovery of cellular division by Rudolph Virchow, in 1855, did nothing to help unanimity.

Since the 1950s, the electron microscope has enabled a more detailed study of the complexity of protozoa. Despite the fact that they have only one cell, these elementary organisms do possess specific systems, for the taking and digesting of food, locomotion, the receiving of information from outside and the co-ordination of behaviour, features which had been thought, until then, to be exclusive to multicellular organisms.

The complexity of protozoa

The discoveries made from the 1950 onwards have contributed to applied research, notably in the struggle against pathogenic (disease causing) protozoa like *Plasmodium falciparum*, which is responsible for malaria. In the mid-1980s they contributed to the perfecting of a vaccine against malaria.

Pulsars

Hewish and Bell, 1967

Lost hearts

The discovery of pulsating stars,
'celestial clocks', is an important
date in astrophysics.

In 1967 the Cambridge Observatory had at its disposal a new type of radio telescope, consisting of 2048 dipole antennae arranged on a surface of 20 square km. It had been designed and set up to establish a new map of the cosmos describing the quasars, which had been discovered a few years before (see p. 173), and classical radio galaxies, by taking as a basis a phenomenon called inter-planetary scintillation; this is a very rapid variation of radio intensity caused by the clouds of ionizing gas which characterize quasars.

The first pulsar

An inventory of all the scintillation sources was in the process of being put together when two young radio astronomers, Ann Hewish and S.J. Bell, discovered a source which was scintillating, though weak, and which had the strange quality of not being always perceptible on the radio surveys of the region where it was to be found. At first the researchers took it for stray interference,

Super-dense matter

One of the most widespread hypotheses on the structure of pulsars suggests a more or less rigid casing or 'shell', inside which a super-dense material is to be found, at a very low temperature and therefore super-conductive. This theory is based on studies of the rotation of celestial bodies.

but close analysis revealed, on 28 November 1967, the presence of a pulsating emission. The result was confirmed a few days later and after verification the observation of this celestial body, the nature of which was unknown, was published. The hypothesis put forward was that it was a star in the process of inward collapse, on the way to becoming a black hole (see p. 29) or perhaps a neutron star. In fact Hewish and Bell had just discovered the first pulsar, so called from a contraction of the words pulsating radio star.

All over the world in radio observatories, where it was rightly supposed that there could be no such thing as a unique phenom-enon in the sky, there began the search for other pulsars, and indeed some were discovered. By 1986, 55 had been recorded, of which 49 came in the first two years after the discovery of the first. There will certainly be others who pulse so rapidly as to make them difficult to detect.

A regular pulse

The exceptional character of pulsars comes not only from the pulsations of their radio emissions, but also from the regularity of the pulsations, which has earned them the nickname 'celestial clocks'. In most cases these pulsations have periods of from 1/2 second to 1 second, although ultra-rapid ones have been traced emitting at periods of 0.033 seconds and others, much slower, with periods of 3.7 seconds. (These figures represent the two extremes.)

However, regularity of emissions is relative, as the example of the first known pulsar bears out, since its pulsation is at times so weak that it cannot be detected. Irregularities of pulsation over long time-scales have also been recorded, and close analysis has also allowed for the detection, even in apparently regular rapid pulsation, of signals lasting a fraction of a millisecond. Finally, a study of the accounts of pulsations in all known pulsars indicates a general tendency towards slowing down. This tendency is obviously weak if account is taken of the fact that it would take some 10 000 years for an average pulsar to double its period and some 2 400 years for a rapid pulsar. One has indeed been found which is so regular that it would doubtless take 100 million years for its period to double.

There lies the enigma

It seems that pulsars are not all alike and, if most of them are visible to the optical telescope, one of the strangest, NP 0532 which is found in the Crab Nebula and which is the most rapid so far known, emits not only visible light, but also X-rays. It is only one of a few sources where this is the case. It is thought that the radio emissions from pulsars are produced by different points in their structures.

Pulsars seem to be more numerous at the equator of the Milky Way and their distribution is therefore close to that of young stars. They are not very distant from us and most would be found some thousands of light years from the solar system; that places them in the interior of our galaxy, which is some 150 000 light years in diameter.

It is in their actual nature that the enigma, and therefore the fascination of pulsars lies. There is no doubt that pulsars are extremely small, as we believe their pulses come from a single emitting spot which flashes past us, rather like a lighthouse. However, due to their extremely small periods, these pulsars must be spinning rapidly. Any large body spinning at that rate would break up under the strain. It is estimated that the average pulsar does not exceed a few thousand kilometers in diameter, that is to say that these mysterious bodies are not much larger than Earth and that some of them are certainly smaller.

The energy radiated by some of them, like the Crab pulsar, is vast, however; from the energy-producing point of view this particular pulsar is more powerful than the Sun. Their nature must therefore be close to that of the stars.

In fact the discovery of the pulsar PSR 0833 in the Nebula Vela X, then of the Crab pulsar, has given much weight to the hypothesis that their nature is stellar. Vela X is a radio nebula similar to those which have been created by explosions of stars (although Vela X does not emit light like these nebulae), and it is known that the Crab nebula is made up of the remains of the star which exploded in 1054 and whose extraordinary fireworks were carefully described by Chinese astronomers. The presence of pulsars in this stellar debris shows then that they are the 'hearts' of extinct stars which continue to beat. The most rapid pulsars would be the result of the most recent star explosions and would therefore be the youngest, such as the Crab pulsar which is probably less than 1000 years old.

The analysis of radio emissions from pulsars has given rise to several hypotheses, none of which was deemed entirely satisfactory in 1986. This analysis in fact revealed that for the recorded periodicities, the density of the matter was somewhere between 100 tonnes and 10 000 tonnes per cubic cm. They could not therefore be white dwarves, as they cannot have densities higher than 100 tonnes per cubic cm, for, above that limit it is proposed that the star's own mass is too large for it to support itself and it collapses to either a black hole or a neutron star. It is possible that pulsars might be composed of neutrons, ie a neutron star; if so, they would be the first celestial bodies of this hypothetical type which have been found.

In fact an atomic model postulates that the inward gravitational collapse which would change an ordinary star into a neutron star would bring about a sudden acceleration in the rotation of the body, with a period close to 1000 rotations per second. Such an acceleration would cause the formation of a great distorted magnetic field which would pull particles towards the star, these particles either becoming trapped in orbit about the star or being hurled away into space at something approaching the speed of light. The ultra-rapid rotation of such a magnetic field from which bundles of particles would shoot out would produce exactly the type of pulsations seen in the pulsars.

This theory accords with the small dimensions attributed to pulsars. The gravitational collapse could — and ought to — reduce a large star to the dimensions of the smallest planet in the solar system.

A promising hypothesis

The only note of apparent discord in the reasoning of the astrophysicists lies in the difference of density between that which was postulated for pulsars and that of the theoretical neutron stars, which is a thousand times greater. The explanation might be that these stars are surrounded by a shell of matter less dense than their core, a plasma of unknown type and probably of a density in the region of the 100–10 000 tonnes per cubic cm proposed for pulsars.

Finally, the hypothesis agrees with the emissions of X-rays produced by the Crab pulsar which are compatible with the 'functioning' of a neutron star.

It is claimed that stars whose collapse has given birth to neutron stars were of medium size, for, if they had been larger, they would have ended up as black holes. Pulsars have therefore greatly enriched astrophysics, by virtue of the analyses and hypotheses their discovery created.

A magnetic hell

The collapse of a star whose diameter suddenly goes from 10 000 to 10 km leads to a growth in intensity of the magnetic field by a factor of 10; this means that, for its small volume, this celestial body induces a magnetic field of disproportionate power. If, meantime, its rotational axis has changed and no longer corresponds to its magnetic axis, the fluctuations of the magnetic field also become huge and the dynamo effect of bodies turning at great speed is sufficient to pull in particles with great force and hurl them into space for considerable distances.

Quasars

Sandage, 1960

At the outer limits of the universe

These 'superstars' constitute one of cosmology's enigmas

In the 1950s astronomers assumed that they had classified all types of celestial body. Therefore the announcement by Allan Sandage of the Hale Observatories (Mount Wilson and Palomar) that he had discovered what was apparently a star of unknown variety in our own Milky Way galaxy caused considerable surprise. The spectrum of this strange object, which was located at the position of a known very powerful radio source, did not show emission lines of hydrogen or helium which would normally have been expected in a star of that colour. Known as 3C48, its designation in the third Cambridge catalogue of radio sources, this was the first optically identified quasi-stellar object or *quasar*.

Early anomalies

The discovery gave rise to confirmations, tests and hypotheses. Scepticism about the identity of the radio and optical source was completely dispelled by the discovery of a second 'star' of the same type, also at the same location as a radio source. This time it was 3C273, which has now been shown to be the nearest quasar to us. In 1962 Cyril Mazard of the Jodrell Bank Radiotelescope in the UK took equipment to Parkes Radio Telescope in Australia to make very accurate measurements of the radio source position. He did this to take advantage of an unusual opportunity; an occultation of the source by the moon when viewed from part of Australia. In such a lunar occultation the moon moves across the line of sight between the observing station and the source. Signals from the source are recorded while the edge of the moon cuts off this line. Thus by using a very accurate clock, a position for the radio source, which is far more accurate than is normally possible with a radio telescope, can be obtained. Mazard had developed this innovative technique at Jodrell Bank in England. At the time of Hazard's measurements radio transmissions were banned for many kilometres around to avoid interference. The positions of the radio source and the strange optical source were found to be identical within a very small margin of error.

Later in 1963 Maarten Schmidt also of the Hale Observatories proposed an explanation which would account for the apparent anomaly in the 'star' compositions. He pointed out that if the spectrum of a normal star is moved to longer wavelengths, then it corresponds to that of the the new object. This 'redshifted' spectrum would arise from the Doppler effect if the quasar were moving away from us at speed. In the case of 3C273 this would be at a speed of 15 per cent of the speed of light or 50 000 km per second.

Subsequently astronomers studied known radio sources and found many other quasars all of which had large redshifts some indicating speeds of recession of 80 per cent of light speed. The consequence of these measurements is that if the redshifts are cosmological, that is if the large redshift means for quasars what it means for galaxies then they must be very far outside the Milky Way at great distances. This presents a formidable problem. These quasars are sources of radio wave energy. When placed at the distance indicated by the redshift the power generated by quasars was greater than which was possible by any mechanism in acceptance at that time. The 'quasar engines' were a source of power without an explanation. There arose the redshift controversy with arguments advanced that quasar redshifts are not distance indicators and that the quasars are much closer so no special 'quasar engine' needs to be invoked. One such arguments is an appeal to the 'gravitational redshift' which would arise if the light were emitted very close to a black hole. However there is difficulty in producing, by this means, redshifts of the size discussed.

Later developments

Quasars' spectra were studied intensively in the 1970s and 1980s, they have both emission lines and absorption lines. The emission lines are unlike those of any star resembling more the spectra of light from a class of galaxies known as Seyfert galaxies. The centres of Seyfert galaxies contain clouds of gas moving at velocities of thousands of kilometres per second.

Another discovery is that the light output of some quasars varies often on a timescale of days. This can only be so if such quasars are a few light days across, this is small in astronomical terms, the size of the orbit of Pluto.

A picture of a quasar is built up from these observations. The average quasar is brighter than 250 billion suns, all this energy is generated in an area smaller than the distance between the sun and the nearest star, a very small distance. Fast moving clouds of gas are close to the centre of this quasar which lives within a surrounding galaxy. The power source or quasar engine is a massive black hole the gravitational pull of which is the centre of the high velocity clouds and which sucks in material of the galaxy to generate the enormous power.

Of the various theoretical models for a quasar this massive black hole one is the most plausible. We can conceive of black holes forming in every large galaxy. Such black holes become activated when they accrete gas or stars. Seyfert galaxy behaviour is a quiet version of quasar activity. The centre of our own galaxy displays Seyfert galaxy-like behaviour at a much lower power level.

Quasi-crystals

Esclangon, 1902; Schechtman, Blech, Cahn and Gratias, 1984

The forbidden structure

Crystals whose structure is not strictly periodic are a challenge to classical crystallography.

In crystallography, the branch of physics concerning solids, two types of structure are classically defined: one is amorphous, the other crystalline. An amorphous structure, of which glass is the classic example, shows itself as a chaotic pile of identical components or substructures. A crystalline structure, on the other hand, shows itself in the form of a strictly periodic repetition of components; quartz is a typical example.

Another aspect is that in crystals an order can be distinguished over long distances because of the rigorously periodic organization of the components, whilst in amorphous structures the figures follow lines which are broken haphazardly and the order is only discernible over short distances.

Diffraction also illustrates this difference: the image produced by a beam of particles — photons, electrons, neutrons — when it strikes an amorphous crystal comprises a central point of impact, corresponding to the undeviated particles, surrounded by rings which correspond to the mean distances between atoms; but when the beam strikes a crystal, the perfect periodicity of the atomic structures brings about regular diffractions of the particles into beams whose deviations from the incoming beam are characteristic of the structure of the crystal. The image resembles that obtained from a grating covered by regularly spaced points and is called Bragg diffraction.

Five-fold symmetry

Crystalline arrangements obey complex laws defined by the following general principle: the components, which are called 'lattices', are perfectly arranged one against the other leaving no spaces in between. The lattices are comparable to polyhedrons which could be piled one on top of the other in the same order and without spaces. Such polyhedrons only exist in limited numbers and, as anyone can discover by trying to put together two-dimensional polygons cut out of cardboard, it is impossible to place pentagons side by side. We can allow that a crystalline lattice be composed of pentagonal elements but not that it be pen-

The memory of matter

In theory, the discovery of quasi-crystals introduced the concept of 'memory' into the constitution of matter, which means that the atoms can repeat a structure according to what are called incommensurable periods between them — that is, periods whose connection is an irrational number. This memory is the function of autocorrelation of the system. The concept gains from its close connection with present mathematical attempts at the interpretation of chaotic states. In practice quasi-crystals have opened the way for the perfecting of extremely resistant alloys, because they have a much greater resistance to dislocation than ordinary metals, by the very fact of their lack of periodicity. Several of these quasi-crystals have been produced, notably in France, by Péchiney.

tagonal itself. This has led to the formal exclusion of structures with five-fold symmetry from the field of crystalline networks.

In 1984 the research of new materials led to the successful production of an aluminium and manganese alloy in special conditions. (It is chilled in its liquid state and cooled ultra-rapidly at the rate of 1 million degrees per second.) As is normal for new materials this one was submitted for crystallographical analysis. It was then that the Israeli Schechtman, then his colleague Blech, of the Technical Institute in Haifa, followed by their colleague Cahn, an American from the National Bureau of Standards of Gaithersburg, and Gratias a Frenchman from the Centre d'études de chimie metallurgique du CNRS, made an extraordinary discovery: on diffraction the micronodules in the alloy, which appeared in the form of very brittle metallic ribbons, revealed a crystalline structure and also an arrangement in five-fold symmetry.

A quasi-periodicity

Crystalline structure was — and remains — indisputable on the examination of the image of the diffraction of electrons; it appears as a regular grid and not like the image in the form of rings centred around a point of impact. Yet the closest analysis obstinately revealed an icosahedric struc-

ture, that of a polyhedron with 20 faces comprising the forbidden arrangements of the pattern of five. Many teams from various parts of the world have verified this challenge to the laws of crystallography. There was nothing for it but to return to relatively ancient theories which foresaw the possibility of the existence of crystals, or more exactly of quasi-crystals, involving a quasi-periodicity.

The first to postulate this quasi-periodicity was the French mathematician Esclangon in 1902. The concept was explored more deeply from the mathematical point of view by the Americans Bohr and Besicovitch, according to the principle which may be very briefly summarized as follows: periodicity, which defines the crystal, consists of the placing one on another a series of identical objects, one of which, at the end of a given time (that of the period), finds itself in a position exactly parallel to the first. A rough idea of this can be gained by imagining a series of steps in a spiral staircase, one of which would be situated, after a certain number of steps had been taken, exactly parallel to the step at the starting point. Bohr and Besicovitch introduced the notion of an infinite number of such steps. Applied to crystallography, this concept means that crystals can be imagined with infinite lattices and no longer solely with long periods. This is the currently held concept of quasi-crystals.

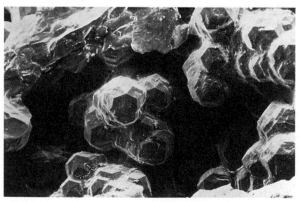

An incommensurable material

These pentahedric crystals were formerly judged to be 'impossible'. They were, however, discovered and named 'incommensurable materials' because their unitary structure is theoretically infinite.

Quinoline

Runge, 1834; Gerhardt, 1837

The ancestor of antibiotics

This chemical base has considerably enriched industrial chemistry and pharmocology.

To understand the discovery of quinoline we have to step back into the last century and experience the atmosphere of the research laboratories of that time. Modern chemistry had existed for only a few decades and every conceivable thing was being distilled, broken down, heated and mixed together to see 'what would turn up'. This apparently erratic practice was reminiscent of that of the search for gold and this is how it was when the German Friedrich Friedlieb Runge, Professor of Chemistry at Breslau, distilled coal tar in 1834. He obtained a colourless liquid with a strong smell, soluble in 16 parts of cold water. Oddly enough, Runge, who was otherwise of perceptive and enquiring mind — we are indebted to him, among others, for the process of extracting sugar from beet (see p. 195), did not analyse this liquid, nor see any particular interest in it; he did not even give it a name.

It was the Frenchman Gerhardt, the great 'organizer' of organic chemistry, who, three years later, realized quinoline's full potential. He had rediscovered it by distilling it from quinine.

From stag's blood to industrial chemistry

It was Gerhardt who observed that it was found in an old remedy, Dippel oil, a distillation of stag's blood known to apothecaries for a century. It was in pharmacology that its properties were to be first put to the test. From it were extracted quinoline tartrate, a white crystalline powder used against fever and infections; quinoline salicylate and sulphoralicylate, which have the same properties; oxyquinoline and orthoxyquinoline, powerful antiseptics which were to preceding centuries what antibiotics are to ours; and, later on, chloroquine and amodiaquine, synthetic quinolines used in the treatment of malaria. Quinoline would also be used as a base for numerous amoebicides such as chiniofon and iodochlorhydroxiquine.

From the 19th century onwards, the treatment of many dermatological, urinary, gynaecological and intestinal problems has been modified by the introduction into pharmacy of quinoline derivatives.

In industrial chemistry too, many colorants have come out of it since the end of the last century, such as the yellow and red of quinoline, the latter being used in the manufacture of orthochromatic photographic plates (sensitive to green as well as blue light), and the cyanins, blue and green colourings also used in photography, printing and the textile industry.

Quinoline is now prepared synthetically following the process, invented by the Austrian Skraup, which consists of oxidizing with nitrobenzene a mixture of aniline and glycerine in the presence of sulphuric acid.

> **Chemically**
>
> Quinoline is an organic compound based around a ring of carbon atoms (a heterocyclic molecule) made up of the coupling of a benzene nucleus and a pyridine nucleus, possessing one isomer, isoquinoline. (An isomer has the same component atoms but a different structure.) Both derive from naphthalene by substitution of an atom of nitrogen in a CH group. The formula for simple quinoline is C_9H_7N. It was Gerhardt who named it quinoline.

Radio waves

Clerk Maxwell, 1865; Hertz, 1988

Waves travelling at the speed of light

The capacity for isolated bodies to transmit electromagnetic radiation remained a mystery for a long time.

Heinrich Hertz demonstrated the existence of electromagnetic radiation in an experiment set up with the purpose of verifying the prediction of James Clerk Maxwell. Scientific discovery does not always follow the seemingly logical progression from theoretical prediction to experimental verification, but the history of electromagnetic radiation spans the nineteenth century and is just such a case of theory preceding discovery. Electromagnetic radiation was predicted by Maxwell and demonstrated by Hertz.

James Clerk Maxwell is the well known Scottish physicist who, bringing together and building on knowledge of electricity and magnetism, created one of the most important formalisms in physics. In the early nineteenth century, the French physicist Charles Augustin Coulomb demonstrated the inverse square law of electrostatic force: that is, the force of attraction or repulsion between electrically charged objects decreases as the separation between the objects increases in proportion to the square of that distance. This was further developed by the German mathematician Karl Friedrich Gauss using the concept of the electric field. Another Frenchman, André Ampère, showed how a magnetic field is generated near a wire conducting electricity. The Englishman Michael Faraday, and independently, the American Joseph Henry established the corresponding fact that electric currents can be induced by a changing magnetic field. These principles form the basis of Maxwell's

work. In 1865 he published his somewhat unwieldy, but nonetheless brilliant formalism predicting the existence of electromagnetic waves which propagate at the speed of light. It took the genius of the English mathematician Oliver Heaviside to simplify Maxwell's work, through the use of vector calculus notation, into the familiar four equations which are now known as Maxwell's Equations. Maxwell's work was highly regarded even well before the experimental verification of electromagnetic radiation.

The ingenious German experimenter Heinrich Hertz made his historic contribution nearly twenty years later in 1888, some nine years after Maxwell's death. The apparatus of Hertz consisted of an electrostatic spark generator fixed onto a parabolic metal sheet which acted as the transmitting antenna. Much of the energy from such a spark can be seen as a bright flash in the visible spectrum, but Hertz demonstrated the existence of a much longer wavelength of light, 66 cm, when his counterpart receiving apparatus triggered a magnetic dipole. For many years these radio waves were called 'Hertzian Waves', but today Hertz is honoured by having his name as the fundamental unit of frequency.

Beginning in the early nineteenth century work in electricity and magnetism, continuing through the mid 1800s and the work of Maxwell, the history of the discovery of radio climaxed in 1888 with the experiment of Hertz. A very important name to mention, despite his work being subsequent to the actual discovery of electromagnetic radiation, is that of Guglielmo Marconi. It was Marconi who foresaw and developed the use of radio as a means of communication, where Hertz did not. Thus the science of radio waves has many great contributors to thank for its development.

Radio waves from space

Jansky, 1932

Emissions from the infinite

*There exist within the universe
innumerable sources of radio waves,
the emissions from which have given
rise to radio astronomy.*

In 1932 the American radio engineer Karl Jansky, who was investigating the sources of static atmospheric signals, heard, in addition to these, in circumstances caused by a distant storm, a continuous hissing in the loudspeaker connected to his receiver. He listened for a long time to this singular sound and noticed that it reached its maximum intensity every 23 hours 56 min. As a first-class observer, Jansky noted that this maximum point coincided with the passage of the Milky Way above the cluster of his antenna; so he concluded from this that the source of the hissing noise was to be found in the centre of the Milky Way. This was indeed so and was the first known reception of radio waves from space.

Jansky's observation was received with great interest by radio engineers and astronomers; it inspired another engineer, also an American, named Greber, to construct his own antenna with a parabolic reflector 9 metres in diameter to analyse short wavelength radio emissions from the galaxy. The resolving power of his equipment, that is to say the minimal perceptible distance between two sources of transmission, was 12°, which was remarkable at the time. It enabled Greber to complete the first radio map of the Milky Way, which was the starting point for radio astronomy.

Radio exploration

Ten years later, at the height of World War II, British radar systems picked up interference so powerful that the operators took it at first for German jamming; this was in fact the first solar radio emission ever recorded. Subsequently the study of solar emissions allowed astronomers to follow astral eruptions through their radio emissions, leading to a much better understanding of them.

Later on recordings were obtained of radio transmissions from the planets, from their satellites and from the Moon. As these transmissions were linked to the magnetic field, they enabled us to add first-class information to the knowledge already available on the structure of these celestial bodies. What now remained was the richest domain of the radio astronomy, namely the exploration of the distant spaces of the cosmos,

---

Radar, satellites and radio astronomy

It was the great progress made in the field of radar which favoured the expansion of radio astronomy at the end of the Second World War. Radio exploration of the cosmos was for a long time limited to wavelengths contained between the 15 metre and millimetre wavelengths. Then the coming of the space age gave new scope to radio astronomy. Radio waves of a wavelength greater than 15 metres are reflected by the Earth's ionosphere and so it is only possible to pick them up by use of satellites. The first satellite equipped with a radio telescope capable of picking up long-wavelength radio waves was the Soviet Saliout 6 in 1979.

---

on which optical information was not so easily available. In 1945 the Dutch astronomer Van de Hulst had predicted that it would be possible to pick up radio emissions from cold clouds of interstellar hydrogen, on a wavelength of 21 cm. It was a bold prediction, for an atom of hydrogen only produces an electromagnetic photon every 11 000 years, but Van de Hulst calculated that, in a column with a 1 cm^2 cross-section extending from the Earth to the Milky Way and beyond, there would be between 10^{20} and 10^{22} atoms emitting a photon, this being probably enough to generate the emission expected in the 21 cm range. In 1951, several radio telescopes had already been constructed and in several parts of the world the predicted emission was recorded. The emissions were analysed; they revealed that the neutral hydrogen had emanated in a relatively thin layer and in spiral form from almost the centre of the galaxy.

Other discoveries

New discoveries were made in 1963 and 1965: the presence of hydroxyl (OH) emitting on the 18 cm band found in interstellar molecules, particularly in the small clouds near to the centre of the galaxy. The emission found in 1965 in regions such as the Crab Nebula and Nebula NGC 6334 was so intense that it was thought that we had come across a great mystery. It was later understood that the emission was so intense because it was amplified by microwaves (this is what is called the maser effect), but it is still not known what had so stirred up the hydroxyl molecules in the first place.

There was a still more surprising discovery in 1970 when emissions were detected of an interstellar molecule which was none other than formaldehyde, a precursor of an amino acid, glycine, then again of water vapour, probably formed by the single addition of an atom of hydrogen to a molecule of hydroxyl, thus forming H_2O in the form of icy droplets. So radio astronomy had just given rise to a new discipline: interstellar chemistry. This in turn raised a famous hypothesis: life has indeed been able to form in space, beginning with water and amino acids. Certain scholars are prepared to go as far as to suppose that organic molecules formed in space have been able to

unleash epidemics on Earth.

Identifying the supernovae

As one might expect in such a field, discoveries abound, since there is now available a new method of exploring the unknown. This is how, in the 1960s and 1970s, radio astronomers discovered about a hundred supernovae: that is, giant stars which suddenly explode at the end of their lifetime. Supernovae have very specific radio characters, such as a 'cold' radio spectrum, which suggests the existence of a shell of energetic particles ejected by the explosion. Already identified, optically, are three supernovae which have exploded in relatively recent centuries and which have therefore been observed by astronomers. (The explosion of a supernova creates a star which is temporarily extremely bright, sometimes so bright as to be visible during the day. One was in the year 1006, well documented by Chinese astronomers; the others were recorded by Tycho Brahe and Kepler. By analysing their radio characters it has been possible to classify the general type of supernovae and thus to identify others. It is also thanks to radio astronomy that it has been possible to identify the mysterious pulsars (see p. 170) and quasars (see p. 173).

20 000 radio sources

Still in the Milky Way, the area of these discoveries, it has been possible to localize individual nebulae and learn more about their structure; in fact these nebulae, real breeding grounds of young stars, are dense clouds of hydrogen which are ionized by the strong ultraviolet emissions of these stars. The ultraviolet light is absorbed by the hydrogen which releases electrons. It is these free electrons that give rise to the radio emissions which can travel through the cloud to us.

The radio cartography of the cosmos has been enormously enriched in recent years and so we have progressed from around 100 known radio sources in 1951, to more than 20 000 today. The greater part of these radio sources is to be found beyond the Milky Way and is composed of either galaxies or quasars.

Once again exploration did not constitute a simple census but made possible the solving of several phenomena which, considered solely from the optical viewpoint, were incomprehensible. At the beginnings of radio astronomy there was a lot of interest in a very distant double galaxy, Cygnus A, which is distinguished by a very high luminosity, a very great speed of recession (16 000 km per second) and extraordinarily intense radio emission. At first it was supposed that the luminosity and intensity of the radio emission could only be the result of a formidable collision between the two elements of this double galaxy. However, analysis has enabled us to correct this interpretation and discover that there is probably an extremely energetic source which lies between the two bodies of Cygnus, which are in fact two plasma clouds created by this source. It is these two clouds, full of high-speed, free electrons, that generate the radio emission, powered by the central source which created them. It remains clear that Cygnus A is, in many respects, one of the most mysterious objects in the cosmos, if only because of the extraordinary energy of its radio emission. It is just one example of the advances made possible by radio astronomy.

The state of the universe

During the 1950s and 1960s, there was great debate amongst astronomers about the origins of the universe. While it was generally accepted that the universe was expanding, there were two conflicting theories that explained how this expansion arose.

The 'Big Bang' theory proposes that the universe started some 15 000 million years ago with a large explosion at a single point, and has been expanding ever since. The 'Steady State' theory maintains that the universe, while expanding, has always looked the same, rather than starting from a single point. This requires matter to be continously formed to replace that lost as the universe expands.

These two conflicting claims can be checked by looking at the distribution of radio sources found outside our galaxy. As light takes a finite time to travel a given distance, when we look at distant objects we are in fact looking at the universe when it was younger. If the 'Steady State' theory is true and the universe has looked the same at all times then there will be equal numbers of radio sources in all parts of the sky. If not, then there will be a change in numbers of galaxies as we look further away (ie look back in time).

The relevant observations and calculations were completed in Cambridge in the late 1960s and indicated that there were more radio sources nearer to us than further away, thus confirming the 'Big Bang' model.

However, this is far from saying that radio astronomy has explained everything in the universe. An enigma like Cygnus A bears ample witness to the discoveries which remain to be made.

Rare gases

Cavendish, 1783; Ramsay and Rayleigh, 1894; Dorn, 1900 and so on

'Lazy' gases

There exist in nature very small quantities of gas which have the characteristic of being normally inert (unreactive).

The first attempt at precise analysis of the air, carried out in 1783 by the English chemist Henry Cavendish, revealed, besides an approximate 21 per cent of oxygen and an approximate 78 per cent of nitrogen (see p. 139), a residue of about 1 per cent which appeared particularly inert, that is to say unable to react with other elements. On this account, Cavendish was the first discoverer of inert gases.

Six precious gases

Taking up these subjects more than a century later, two other Englishmen, Sir William Ramsay and Lord William Rayleigh, discovered by spectroscopic analysis (see p. 115), an inert gas which they named argon, from the Greek word meaning 'lazy'. We now know that argon represents 0.93 per cent of the air's volume.

The properties of helium

Sometimes used for the inflation of balloons, helium has properties precious in physics, including that of being a superconductor at very low temperatures. It is also used in the mixtures of gas for deep sea diving, in the cooling systems of nuclear power stations, and in techniques of geological dating. Argon, krypton and neon are used in lighting tubes, now referred to as 'neon lights'.

The following year Ramsay and, separately, the Swede Per Theodor Cleve identified in the rare ore cleveite, another inert gas which they named helium because it had already been detected by spectrographic analysis in the Sun, by Janssen and by Lockyer. The spectrographic absorption feature picked out in 1868 by these two researchers being unique to helium and corresponding exactly with that of the gas found in cleveite, it could be deduced that one was dealing with the same gas. This one is only present in the air in very small quantity: say, 0.0005 per cent.

In 1898 Ramsay, spurred on by his previous discoveries, and his fellow countryman Morris William Travers, isolated three more inert gases: neon, krypton and xenon, present in the air in amounts of 0.0018 per cent, 0.0001 per cent and 0.000009 per cent respectively. The last of the inert gases was discovered in 1900 by the German Ernst Dorn in the waste deposits of radium, which is why it is called radon. Its properties in the air have not been determined because they are not constant. It is known that granite rocks give off tiny amounts of this gas.

Monoatomics

Radon apart, these gases exhibit the characteristic of being monoatomic (consisting of single atoms) and having complete electron layers, which means that they tend not to form covalent bonds — hence the name inert gases. In 1962, however, the Canadian Neil Bartlett verified the hypothesis proposed in 1933 by the American Pauling, according to which it would be possible to extract an electron from an atom of xenon: by making platinum hexafluoride act on xenon, Bartlett obtained, in fact, xenon tetrafluoride.

Red corpuscles

Leeuwenhoek, 1675 or 1684

A genius of the microscope

It was with rudimentary methods that the inventor of the microscope was the first to observe red blood cells.

When, in 1661, the Italian anatomist Malpighi completed Harvey's discovery of the circulation of the blood (see p. 41) through his microscopic observations of capillary vessels, he could hardly fail to attract the attention of a genius of microscopy, Antonie van Leeuwenhoek, a Dutch amateur, later ranked among the great scholars. Leeuwenhoek, who had built a great many microscopes, examined blood under a microscope for the first time in 1684 (or perhaps as early as 1675). He discovered round cells, flat at the centre, of which he gave an excellent description, which testified to the truth of the discovery.

> **Erythrocyte**
> An anucleate cell normally the most common formed element in circulating blood, filled with haemoglobin and shaped as a biconcave disc. It is formed in the bone marrow. Its primary object is to transport oxygen round the body.

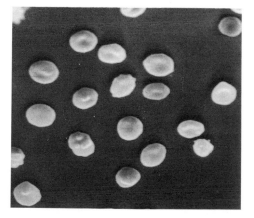

Red blood cells *The disc shape is easily seen.*

Respiration

Priestley, 1771; Lavoisier, 1777–80; Laplace, 1780

Strength drawn from the air

*Understanding of the process of the
assimilation of oxygen by living
beings was the end product of a long
train of discoveries.*

Until the end of the 18th century nothing was known about respiration. Although there had been no shortage of anatomical explorations since the early Middle Ages, it was not even known that oxygen in air played a part until a clergyman, political theoretician, teacher and amateur scholar got involved in experiments with gases: it was Joseph Priestley, dabbler of genius, who has gone down in history for his role in the discovery of oxygen in particular (see p. 145).

It is true that Priestley did not establish all the mechanisms of respiration, which are of course complex and very different in plants and animals; but he did get the research underway and showed first of all that the volume of air decreased by one-fifth during respiration, combustion and putrefaction. In 1771, following his own intuition, Priestley wanted to verify whether or not the air 'used' by animals could be suitable for plants. So he placed a green plant in good condition under a belljar where a mouse had died from asphyxiation: not only did the plant not wither, but it also restored the 'goodness' of the air, since a mouse was able to live there for some time. (In fact the plant had absorbed the carbon dioxide and produced oxygen, the opposite of what occurs in animal respiration.)

Respiration and combustion

This work obviously interested Lavoisier, who was currently carrying out research on gases and had just discovered oxygen (see p. 145). In 1777 he repeated Priestley's experiment, but in his own way. He noted that a bird enclosed in a belljar produced a gas which was absorbed by carbonate of lime (it was carbon dioxide), and that the volume of oxygen in the belljar decreased as respiration went on. As the first of these phenomena only occurred when a living creature was placed under the belljar, and the second phenomenon occurred both when a living creature and a flame were under the jar, Lavoisier deduced, not without some justification, that there was a similarity between combustion and respiration, since both occur thanks to the consumption of oxygen. He set out the basis for understanding the production of animal heat through respiration, a basis which

---

Respiration and fermentation

It is impossible to dissociate the concept of respiration from that of oxidization, that is to say, from chemical processes catalyzed by enzymes. In this regard fermentation (see p. 69) is very close to anaerobic respiration. The methods of cellular use of oxygen continue, moreover, to inspire biological research. Sports medicine is also concerned by anaerobic respiration, as is oncology, since cancer cells make for an increased consumption of oxygen.

---

remains unchanged to this day, although its detail has been considerably enriched.

For Lavoisier, the analogy between respiration and combustion went further than his predecessors had imagined. Combustion being an oxidization, it followed that respiration too is an oxidization. So the scholar considered the phenomenon of respiration in the field of chemistry. In 1780, with the collaboration of another great scholar, Laplace, and using a piece of the ice calorimeter, equipment which they had invented together, Lavoisier verified his theory quantitatively: the respiration of an animal in a belljar brought about a rise in the surrounding temperature, and therefore a release of heat.

Followers and precursors

It was left to Priestley, the Dutchman Ingen Housz and the Swiss Saussure and Senebier to clarify an essential point: the difference between animal and plant respiration. In 1774 Priestley had demonstrated that plants exposed to light produce oxygen, unlike animals. In 1779 Ingen Housz had shown that in darkness, by contrast, they produce carbon dioxide. In 1804 Saussure demonstrated that plants use water and carbon dioxide to increase weight. In 1807, finally, Senebier showed that plants transform carbon dioxide (CO_2) into oxygen, in the course of reactions where the plants absorb the carbon of CO_2 in order to release oxygen.

How respiration gives energy and water

Between 1916 and 1920, Thunberg, developing Wieland's theory, demonstrated that assimilated oxygen picks up hydrogen, which it extracts from organic molecules, to give water. The subsequent breaking up of these molecules releases significant quantities of energy by complex chains of chemical transfers — transfers of electrons in fact.

It was not until after 1950 that it became possible to put together a general theory of animal and plant respiration, both aerobic and anaerobic.

Some relevant discoveries had been made much earlier. For example, the British mystic Fludd had shown in 1620 that a flame burns air and, in 1650 the German Guericke had shown that a candle would go out when placed in a metal sphere where a vacuum had been created. That same year, 1650, the Briton Mayow had already observed the similarity between combustion and animal respiration. In 1727, more than 40 years before Priestley's observations on respiration, his compatriot Hale had already observed that air had 'something' to do with the growth of plants. However, somewhat unjustly, it is the names of Priestley, Lavoisier and Laplace which are most often mentioned in the history of the discovery of respiration.

In any case the knowledge of the physical and chemical basics of animal and plant respiration also demanded that the physiological mechanisms of this function be taken into account. Current notions were vague or simply false. Until Harvey's discovery of circulation of the blood (see p. 41), it was believed that the function of the air that one breathed (for everyone could verify that air was breathed in and out) was to cool the heart in order to produce the 'vital spirit' in the right ventricle (heart chamber). ... The first steps in the right direction were taken when the Englishman Hooke demonstrated that an animal could be kept alive, even without movements of the rib cage, if air was blown into its lungs. That still did not eliminate the notion of the 'vital spirit', which only began to fade after the observation of another British scientist, Lower, in 1672. This anatomist, who had been the first to carry out — with disastrous results — a blood transfusion, noted the difference in colour between arterial and venous blood and explained it, correctly, by the fact that the first had 'absorbed' air. Mayow, mentioned above, had remarkable intuition concerning this, since he suggested that this difference in colour could be due, not to the overall action of the air but to the action of an element of it, which he obviously could not know because oxygen had not yet been discovered. Mayow even went as far as to imagine, but much too tentatively, that respiration consisted of a gaseous exchange between blood and air. All the same he just fell short of discovery proper, for he believed that air gave the body a 'nitro-aerian spirit', although his keen powers of reason had led him to suggest that the air took away vapours emitted by the blood.

Anaerobic respiration

Astonishingly enough, for several decades knowledge of respiration hardly progressed at all. No appreciable discovery in the physiology of respiration occurred before 1849; then Hutchinson studied pulmonary elasticity, barely a discovery. At last, in 1861, the German Ludwig took an interest in variations of gases in the blood during muscular effort. In 1875 the German Pflüger rediscovered a fact already established by Saussure: even when there is no oxygen, oxidizations may continue and end in decomposition. He failed to grasp the real meaning of his discovery, which was not given a reasonable interpretation for another three years, when Pflüger's compatriot Pfeffer spoke of intramolecular respiration. This was respiration that occurred without oxygen, a phenomenon later called anaerobic respiration; it is the opposite of aerobic respiration, which can only occur in the presence of oxygen.

It was only in 1897, in the light of Pasteur's work on fermentation, that the German Bach again raised the level of scientific awareness, which had been pretty sluggish since the time of Lavoisier. Respiration does not bring about direct oxidization of the substratum, for example pulmonary and muscular cells, by oxygen from the air, but a fragmentation of organic molecules at the end of a long series of reactions triggered by enzymes; it is this which releases the energy.

It was well into the 20th century before all the facts and notions about respiration and cellular oxidizations were successfully brought together. From 1912 to 1949 the Germans Wieland, Thunberg, Warburg and Keilin explained — first by a series of contradictory theories, then by facts — the assimilation of oxygen which Lavoisier had merely outlined. The greatest honour went to Warburg, who discovered in 1930 that the molecular oxygen of the air can only be used after being transformed into atomic oxygen by an enzyme containing iron, which picks up the oxygen.

Restriction enzymes

Smith, 1970

A genetic instrument

These 'DNA scissors' have facilitated the miracles of genetic engineering.

There exist in chromosomes genes which are called repressors because they block the cell's constant phenomena or processes of adaptation; and other genes which are called effectors because they, on the contrary, set in motion the processes of adaptation. These genes are meant to maintain the equilibrium of the cell in changing conditions. This action is exerted by the series of enzymes whose synthesis is governed by the genes.

In 1970, the American Hamilton Smith completed a famous experiment. He incubated *Haemophilus influenzae* bacteria, responsible for a variety of influenzas, with the DNA of a variety of viruses called bacteriophages, which he had previously rendered radioactive in order to be able to observe them better. The bacteriophage viruses are capable of infecting bacteria (hence their name) by introducing their own DNA into that of the bacterium (see p. 52). Smith made the following observation: the DNA of the virus dissipated rapidly. He postulated that it could be a question of a restrictive phenomenon, provoked by the bacterium which was serving as a temporary host for the virus and which therefore must be of enzymatic nature. He tried to retrieve the enzyme responsible by refining extracts of the bacterium and he found it. Thanks to this enzyme he was able to cut out DNA, but only DNA of a type foreign to the enzyme. This was the discovery of restriction enzymes, a major event in the history of molecular biology.

Widely used in genetic engineering laboratories, restriction enzymes (of which there is a wide variety) serve to cut a given DNA at a precise place in the sequence of nucleotides which constitute it. The dozens of restriction enzymes now at our disposal are derived from different bacteria and operate at different incision points of other DNAs.

Genetic grafting

The benefit of restriction enzymes is that they enable combinations of DNA to be achieved for experimental ends. The most recent use of these enzymes consists of isolating a segment of DNA from a chosen cell which regulates the synthesis of a given substance, (interferon, for example), then introducing at a given point the DNA of a host-cell of another type (a bacterium, for example) and inserting the particular segment there. The host-cell then begins to produce the substance synthesized by the segment of DNA which has been grafted on to it. This is 'genetic grafting'. The host-cell most often chosen now is *Escherichia coli*, in which large quantities of precious interferon, a powerful anti-infection substance, have successfully been produced.

Eleven fragments

The American Daniel Nathans succeeded in cutting the DNA of a virus into 11 different fragments whose roles he identified. Nathans, Smith and the Swiss Werner Aber were jointly awarded the Nobel Prize for medicine in 1978.

Harnessing bacteria

Restriction enzymes are the tools which have facilitated the mass production of therapeutic substances formerly available only in small quantities and at great cost; for example, interferon cost £25,000 per gram at the end of the 1960s.

Rings of Saturn

Huygens, 1659

The rings in the sky

These were the first planetary rings
observed in the solar system.

In July 1610, armed with a telescope with a magnification of $\times 32$ (see p. 22), Galileo observed the planet Saturn quite clearly; he noticed on either side of it protruberances, the nature of which he did not understand. These were the rings of the planet, but Galileo whose telescope did not give him a clear view, concluded that Saturn was flanked by two companion planets. In 1613 Galileo once again observed Saturn and found to his surprise that the 'companions' had disappeared. In fact the rings edge-on had become invisible to him.

Thousands of rings

In 1655, equipped with a more powerful telescope than Galileo's, the Dutchman Huygens began work in this field observing Saturn regularly. In 1659 he concluded that the planet was encircled by a ring which did not come into contact with its surface at any point. Did he have faith in his dis-

covery? The fact is that he published it in the form of anagrams, which were deciphered and then rejected as absurd.

Several years later, however, Huygens' idea had gained ground and was no longer contested. Although Galileo was the first to see the rings of Saturn, Huygens was actually the first to identify them and therefore the credit must be given to him.

In 1675, the Italian Cassini observed from Paris that the ring included an important circular gap still called Cassini's division. It was supposed at the time that there were two rings, but in 1850 the Americans Bond and Bond and the Englishman Dawes discovered independently a third inner semi-transparent ring which was nicknamed the 'Crepe Ring.' Following that more rings were discovered, and with the flight of the Voyager spacecraft in 1980 and 1981, several thousands in fact, some very wide, others extremely narrow. The rings were neither completely solid, nor liquid, but formed from masses of fragmented matter including huge lumps, dust grains and also very probably ice particles.

During the 1970s and 1980s rings have also been discovered around Jupiter, Uranus and Neptune. Several theories have been put forward to explain the rings' origin. On the whole it seems that they were created by the disintegration of moons when these got too near to the mother planet and underwent gravitational tide effects. It is as if our Moon were to descend to an altitude of less than 18 000 km from Earth; it would then disintegrate into pieces whose approximate size could be calculated as being, say, 200 km in diameter.

The Roche Limit

The lowest orbit below which a moon disintegrates is called the Roche limit. From this we can predict that in some hundred thousand years, *Triton, a satellite of Neptune* and one of the largest moons in the solar system, may disintegrate and form a ring round the planet. Its retrograde (backwards) movement is bringing it ever closer to the planet around which the American spacecraft Voyager discovered a thin ring.

Rubber

Herrera Tordesilla, 1615

A ball that went a long way

Only after three and a half centuries
of observation did Europeans finally
make use of the sap of the hevea tree.

Rubber has been known in Europe since the discovery of the new world. The first European to describe the elastic gum was Pietro Martyre d'Anghiera, chaplain to the court of Ferdinand and Isabella of Spain. He described an Aztec game played with balls 'made of the juice of certaine herbe (which) being stricken upon the ground but softly (rebounded) incredibly into the ayer.'

At the time of his second voyage to the New World, Christopher Columbus saw Indians in Haiti who played with balls made from the sap of a tree but he was not much interested. In 1615 the Spaniard, Herrera Tordesilla, finally paid attention to this mysterious substance. He could be said to be the original European discoverer of rubber. However, another century passed before the Frenchman, La Condamine, who was sent to South America by the Academy of Sciences, mentioned the hevea tree. This was between 1736 and 1744. The greatest progress in the easy manipulation of rubber came at the beginning of the 19th century from the experiments of a Scottish chemist, Charles Macintosh, and an English inventor, Thomas Hancock who were searching for a solvent to mix with the rubber from South America. Macintosh rediscovered coal-tar naphtha and found it to be an effective solvent. He placed the solution of naphtha and rubber between two fabrics thereby avoiding the brittle surfaces common in garments treated with rubber. Manufacture of these double textured, waterproofed cloaks known as mackintoshes proved very successful. Hancock meanwhile turned to the production of elastic thread and invented a masticator; the heat of friction welded the scraps of rubber together, which could then be applied in further manufacture. However although Macintosh and Hancock resolved the initial problem of manipulating the raw material, no rapid extension of the use of the rubber could be made until the effects of temperature on natural rubber had been overcome (see vulcanization p. 221).

The first rubber tyre was patented in 1848 by the English veterinary surgeon Dunlop, and in 1876 the Englishman Wickham planted heveas in London. Once they were established he exported them to Malaysia and Ceylon (Sri Lanka) where they were used to found great rubber plantations.

The Indians

The Indians used the hevea sap to coat their shoes and their earthen bottles, which then became watertight. This was rubber (which they called *kahuzu*).

Selenium

Berzelius, 1818

An excellent detector

*This rare element plays a major role
in electronics.*

In 1817 chemists found a reddish deposit on the floor of a room in which sulphur had been processed, extracted as a by-product from a copper mine. At first it was thought to be tellurium, a rare element identified in 1798 by the German Klaproth. The Swede Berzelius analysed this deposit and established that it was a specific element, related to sulphur. As Klaproth had named tellurium from the Latin word *tellus*, 'earth', Berzelius called the new element by a name similarly inspired, from the Greek word *selene*, 'moon'.

Following this, it was discovered that selenium may exist in several forms, of which one is vitreous (non-crystalline), another metalloid (with non-crystalline and crystalline properties), and a third metallic. Selenium was to be the object of much more important discoveries, however. The first, in 1873, was that of its photoconductivity and then, in 1877, that of its photovoltaism (see p. 201) which was to lead, several decades later, to the realization of the first attempts at television. Meanwhile, research was to lead to the perfecting of many technological products, such as photoelectric

and photovoltaic cells, whose present-day uses and derivatives are innumerable, from colorimeters to civil and military detection equipment. Selenium also has the quality of being sensitive to the surrounding temperature which, like light, increases its conductivity, making it a thermal as well as optical detector. In its vitreous form, which is amorphous (not crystalline), selenium exhibits no conductivity at all, however, and can be used as an insulator.

A biological role

Poisonings, often serious, recorded in the early years of the selenium industry caused doctors to take an interest in its biological effects. Its strong toxicity was discovered in 1842. From 1929 onwards extensive research has helped to establish that the toxicity of numerous plants is due to the fact that they grow in soils rich in selenium–iron compounds. However, it has also been discovered that many plants need selenium for their growth. In the 1950s Schwarz and Foltz, for example, also discovered that infinitesimal quantities of selenium give protection against necrosis (cell death) of the liver. Since then, other work had allowed for the classification of selenium as one of the elements necessary to cellular equilibrium, notably that of the renal and cerebral tissues.

Selenium and agriculture

Between 10 and 20 parts per million of selenium in a food are enough to cause poisoning. As selenium is now absorbed by numerous vegetables, and by cereals, many agricultural areas have been abandoned, notably in the United States, because they contained too much selenium in their soils. Paradoxically, arsenic is an antidote for selenium. Plants rich in selenium are characterized by a strong smell of garlic.

Selenium and uranium

The presence of selenium in a soil can also indicate that of uranium, the two minerals being associated in certain types of deposits.

Special effects in cinema

Méliès, 1894

The origin of the cinema of the fantastic

*Many special effects in fantasy films
derive from one mechanical accident.*

The beginning of the special effects of which the makers of fantasy films are so fond might have been taken for an invention, had it not been dictated by chance. We will therefore treat it as a discovery.

Méliès was the first of all the directors of fantasy films.

'Would you like to know', he asked in his *Vues cinématographiques*, 'how I first got the idea of applying effects in cinematography? Very simply, here it is. A snarl-up in the equipment I was using at the beginning (rudimentary equipment in which the film often tore or became caught and refused to roll) produced an unexpected effect one day when I was routinely filming the Place de l'Opéra; it took about a minute to free the film and get the equipment going again. During that minute, the passers-by, buses and cars, had moved of course. On projecting the strip of film, patched together at the point where it had broken, I suddenly saw a Madeleine–Bastille bus changed into a hearse and men changed into women.'

Méliès, who until then had used theatrical tricks, trapdoors, smoke and so on, now had a vision of specifically cinematographic effects, called 'substitution' or 'freeze effects'. To begin with, these allowed him to achieve simple substitutions; then, by a series of improvements, he achieved dissolves, appearances by superimposing on previously exposed white backgrounds (then held to be impossible), metamorphoses and so on. In the end Méliès was able to have an actor play 10 roles simultaneously, by re-exposing the film 10 times.

Méliès and E.T.

Although originally helped by pure mechanical accident, Méliès had the genius to exploit it using his own inspiration. Recourse to mechanical effects, conjuring tricks and optical illusion helped him to lay the foundations for the technique of special effects which has developed — still on those same principles — to produce masterpieces of cinematic illusion like *Star Wars* or *E.T.* The technical expertise of specialists like the American Lucas has, even today, barely altered the principles on which Méliès based his work such as *Le manoir du diable* or *Le diable au couvent*.

Stop Motion

This technique involves filming a model using one frame at a time; it was first used by Willis O'Brien in 1913 in the film *The Dinosaur and the Missing Link*. He developed the technique for *King Kong* (1933) using transparencies to overlay pictures of the puppet on the actual film.

Speed of recession of galaxies

Slipher, 1912

Flights towards infinity

*The galaxies are moving away from
each other as the universe expands.*

At the beginning of the 20th century important technical progress was made in spectrography (the use of a spectroscope designed for a wide range of frequencies), thanks to more sensitive equipment and to very wide photographic lenses. It became possible to make much more precise observations of faint celestial formations like extragalactic nebulae. So in 1912 the American astronomer Vesto Melvin Slipher came to make a fundamental discovery in cosmology: galaxies are distended objects and are travelling at great speed. Thus the Andromeda nebula was found to be receding at the radial speed (the speed calculated along the observer's line of sight) of 300 km per second.

Speed of recession

and speed of approach

The speeds of recession of galaxies are not all the same. By 1929, in fact, 46 of them had been recorded, the highest of which, measured by Slipher, was of 1800 km per second; in 1931 a speed of 19 600 km per second was registered for the star cluster Leo; and, since 1960, speeds reaching and surpassing 100 000 km per second have been recorded. Most of the galaxies were found to be receding from us; however, some galaxies

have been observed that are moving towards us: for example, Messier 31 in the constellation Andromeda, which is approaching us at the speed of 300 km per second. However, all the galaxies that are moving towards us are in fact very close by.

In physical cosmology the discovery of the recession of galaxies was only assimilated after several decades: it was to lead to a theory of the origin of the universe, which maintains that the universe began some 15 billion years ago with a massive explosion (the 'Big Bang'), since then the universe has been in constant expansion. This theory is supported by the fact virtually all galaxies are receding from us at speeds proportional to their distance (see p. 56).

From one notion to the other
In theoretical cosmology the recession of galaxies opened up to question the notion of the static universe, which, at the time, was automatically assumed. Indeed, Einstein, on formulating his general theory of relativity and finding that it naturally gave rise to an expanding universe, introduced an arbitrary constant, the cosmological constant, in order to reconcile his theory with the notion of a static universe. He was later to confess that this was the greatest error of his life.

Spermatozoa

Leeuwenhoek, 1677

They were mistaken for parasites

The identification of the male fertilizing cells was not an immediate aid to the understanding of fertilization.

The story of spermatozoa (like many other stories in science) wonderfully illustrates the damaging effects of ideology in the interpretation of reality. The man who discovered them, the Dutchman Antonie Van Leeuwenhoek, also the discoverer of bacteria (see p. 102), owed his luck to the observation through a microscope of different seminal fluids. With his objective mind, Leeuwenhoek merely described them as exactly as possible. It surprised many people, but no-one looked into the matter seriously. On the contrary, the following year, his compatriot Hartsoeker published engravings to represent the cells observed by Leeuwenhoek: these showed tiny humans or homunculi, evidently preformed. Yet Hartsoeker had observed the spermatozoa through a microscope!

Preformation or epigenesis

Until the beginning of the 19th century and until considerable progress had been made in microscopy, two theories continued to divide scholars on the formation of the embryo. One was the theory of preformation, according to which there was, in the sperm, a completely formed individual which, in the course of gestation, increased in matter until reaching the size of the newborn, The other was the theory of epigenesis, which claimed that the egg only developed during the course of a series of stages. This was close to what really does happen, but no-one understood how a formless substance could become organized into a miniature human being.

One of the most extraordinary paradoxes in this area is that in 1773 the great Italian physicist Spallanzani, who was the first to succeed in a laboratory fertilization of frogs' eggs with semen, maintained that the spermatozoa were merely parasites in the semen. Spallanzani put forward a theory as fantastic as it was complex, which consisted of a variation on the ideas of preformation. According to him every individual was present and preformed in the egg (ovum) of female animals, and the sperm only played a limited role as liquid nourishment.

Several years before, however, in 1768, the work of the German Wolff on the development of the chicken embryo had added grist to the mill of the thesis of epigenesis. Then in 1845 the German Baer founded embryology by showing the existence of three germinal layers in the embryo. The preformation theories increasingly lost ground until Hertwig dealt them a fatal blow in 1875 by showing that the spermatozoon and the ovum are actual cells whose fusion is necessary to fertilization (see p. 70). At a stroke embryology entered into the realms of science.

> **Influence of a theory**
> Spallanzani's theory prevailed for decades and, influenced by religious considerations, some of its supporters even maintained that sexual intercourse was not necessary for procreation
> ...

Spiral galaxies (Rotation of)

Herschel, 1800; Baade, Morgan, Oort and Van de Hulst, 1950–1

The galactic merry-go-round

The discovery of the rotations of galaxies.

When observers first explored the skies by telescope they saw objects which were not point sources but had structure. They called them nebulae. The Englishman Herschel working around 1800 saw some of these and realized that they were spiral; he was able to describe them and draw them. At the beginning of the 20th century astronomers were able to study photographs of nebulae and galaxies which clearly revealed their shape. A study of the shape led to specu-lation that some of the galaxies and nebulae shown in the photographs were spiral, but the photographs at that time were insufficiently detailed to allow confirmation. Between 1925 and 1930 the Dutchman Oort, studying the apparent motion of stars in our own galaxy, the Milky Way, discovered that the Milky Way itself was turning.

> **Celestial calculations**
> The discovery of the rotatory movement of the spiral nebulae enabled the calculation of the mass of galaxies and the dynamics of celestial objects in space.

M83 is a large, almost face-on spiral galaxy perhaps 10 million light years away. The central portions, in common with many galaxies, have a bar-like structure from which the strong, complex, spiral arms extend.

Sugarbeet

Marggraf, 1747

A product of the British blockade

*The idea of getting sugar from beet
at first seemed a mere oddity.*

Since about the 8th century sugar as a food-stuff had been obtained in Europe and the East from sugar cane, which was grown notably in the South of France and in Spain. Then Columbus introduced it into the New World in 1493 and the success of sugar cane growing in Santo Domingo led colonists to create plantations in the Caribbean (which also encouraged slavery). No other plant sources of sugar were then known.

In 1747 the German apothecary Andreas Marggraf, who obviously knew about beet as a laxative, wondered if it would be possible to extract sugar from it, seemingly a pointless idea since there was no sugar shortage in Germany. So he ground up some beets, probably of the Silesian or Magdeburg variety, and obtained a juice from them which he filtered and evaporated through heating; the crystalline residue had a taste comparable to that of sugar cane. (The apothecary subsequently acquired some fame by persisting in his adherence to the theory of phlogistics — see p. 143.)

No-one was interested in his discovery, except for one of his pupils, a German of French origin, Franz Karl Achard, who planted beet on his land in order to find out which variety gave the best return. He succeeded in interesting Friedrich Wilhelm III, King of Prussia, in his research and created in 1802 the first pilot sugar beet refinery at Cunern in Silesia. In 1810 Marggraf's 'oddity' had become an industrial reality. The inventor, however, had died in 1782.

A worldwide destiny

In 1811 the British blockade deprived France of its supplies of sugar cane from the Antilles, and production in the South of France was insufficient. That same year the financier Delessert set up a small refinery at Passy and obtained sugar which was perfectly crystallized and marketable; this earned him the title of Baron and the *Légion d'honneur*. Napoleon immediately had 40 sugar beet refineries created throughout France.

The restoration of maritime trade with the Antilles threatened the sugar beet industry with collapse for a time, but its economic advantages had become too obvious to be ignored and the West as a whole adopted sugar beet as well as sugar cane.

In 1838 the Americans Church and Child founded the first sugar beet refinery in the United States. The pulp and treacle of sugar beet still serves, after treatment, as animal feed. The treacle is also used in the production of citric acid, by fermentation.

A sedative and a gem
Until the end of the 18th century sugar was considered as a luxury commodity and its use then bore no relation to its use today. Apothecaries and doctors prescribed it as a sedative — the practice of giving insomniacs a glass of sugared water was carried on in some countries until the beginning of the 20th century — and sugar crystals were comparable to precious stones: they were even mentioned in the dowries of princesses. ...

Sunspots

Galileo, Fabricius, Scheiner and Harriot, 1610–11

Regulation of the human temperament?

*The existence of dark spots on the
face of the sun was the first
indication of its intense activity.*

Within the space of a few months, more or less simultaneously, the Italian Galileo, the Dutchman Fabricius, the German Scheiner and the Briton Harriot made the first discoveries of the spots on the surface of the Sun, between 1610 and 1611. The discovery was made possible thanks to refinements in the astronomical telescope (see p. 22). However, the only one who sensed that the phenomenon was linked to the nature of the Sun was Galileo.

The observation was more or less left at that for two centuries until the German amateur astronomer Schwabe, who had been patiently observing the spots for 33 years, discovered that the mean number of spots was subject to cycles of about 10 years. In 1852 the Swiss Wolf published a new and precise calculation: these cycles were of 11.2 years and the possibility of an 80-year-long cycle was mentioned.

Magnetic anomalies

In 1858 came a new discovery: the spots appeared at the solar latitude of 30°, and as the cycle went along they got nearer to the equator, ending at the latitude of 8°. Moreover, the spots disappeared at the end of one solar rotation, which was 27 days.

Since 1834 magnetic observatories had existed, like the one which Gauss had established in Göttingen; in 1857 it was discovered there that magnetic storms on Earth coincided with the peaks of the sunspot cycles. The astronomical observatories, which, like those of Zurich and Greenwich, provided daily surveillance of the Sun, and the magnetic observatories agreed to collaborate. In 1904 a violent magnetic storm coincided with the passage of large spots at the central solar meridian, confirming the connection.

The nature of the spots

At the end of the 19th century the magnetic anomalies were the only characteristic which could be attributed to the spots. In 1891 the American Hale invented spectroheliography, a branch of spectrography (see p. 196) dedicated to the study of the Sun, so it became possible to analyse in more detail certain observations made since 1870, notably the formation of gaseous flows around the edges of the spots. In 1909 it was discovered that these flows began at the centre of the spots, which were relatively cold, and moved out towards the edges at a speed of 2 km per second. Between 1914 and 1924 Hale tested and modified the theory of the German Kirchhoffer, according to which the Sun was surrounded by a

The magnetic intensity of the spots

A small sunspot has a magnetic field of 500 gauss (a gauss being the unit of measurement of the strength of a magnetic field); a large spot, a field of 4000 gauss. In comparison, the average solar magnetic field varies between 20 and 50 gauss, and the Earth's magnetic field is around 1 gauss.

colder gaseous cloud: at a higher altitude than that of the centrifugal gaseous flows, Hale observed similar centripetal flows. The spots were revealed to be intense centres of activity.

It is now known that the spots are made up of relatively cold troughs of low pressure — they only appear dark by contrast — some of which may be several times larger than the Earth. They seem to be caused by irregularities in the convection of the Sun's internal energy, from the centre to the surface: a 'thaw' is produced on the magnetic force lines, which escape from the very intense field by forming some sort of 'bubble'. It is this kind of field which causes magnetic disturbances on Earth, after the fashion of the solar eruptions.

Exact cycles

Established for the first time in detail by Wolf, these sunspot cycles were shown to be much more variable than had been supposed earlier; they have been known to have minimum periods of 8 years and maximum periods of 16 years, the figure of 11.2 years

---

The Sun King and the Sun

As sunspots have been observed continuously since 1610, their history has been known for about three and a half centuries. It is interesting to note that the longest period of solar calm coincides almost exactly with the reign of Louis XIV of France, called the Sun King, which lasted from 1643 to 1715, the Sun's calm period being from 1640 to 1716.

The Sun

The Sun is the central object of a solar system and the nearest star to earth. The source of its energy is nuclear reactions in the central core (temperature 15 million K). Every second it annihilates 5 million tonnes of matter, to release 30×10^{26} watts of energy.

---

merely representing an average. Also, long cycles of low amplitude, of 80 years, have been detected.

Earthly correlations

It has been possible to verify that the spots and solar eruptions provoke disturbances — fadings — in shortwave transmission; even were it viewed from this angle alone, the discovery of sunspots would be important in the whole area of radiocommunications.

Attempts have also been made to find out whether a correlation exists between the cycles of the spots (as well as those of eruptions which are, moreover, concordant) and certain other earthly phenomena, by reason of biological mechanisms being sensitive to the electromagnetic field. A lot of research has gone on in this area. A pioneering role was played by the Russian Tchijevsky who, in 1938, published a series of statistical correlations between the solar cycles and the cycles of morbidity (illness) in the world. Tchijevsky even found a correlation between the excitability of crowds and the cycles. In 1969 the American Cogan and, in 1971, his compatriot Dewey found correlations between solar cycles and the economic and financial fluctuations in the West. (The world crisis of 1929, as well as the war of 1939, closely followed the ends of cycles, so perhaps earlier correlations also ought to be taken with a pinch of salt.)

It is true that fluctuations in the Earth's magnetic field closely follow those of the Sun's magnetic field, and that they influence many biological phenomena, in a way as yet only dimly understood. For the moment these correlations do not imply any significant causal link. If there were to be such a link, which cannot be ruled out, it is clear that the discovery of sunspots and the knowledge of their effects and mechanisms would assume for science an importance considerably greater than that which they already have in astrophysics and electromagnetism.

Superconductivity

Onnes, 1911

Changes brought on by the cold

At very low temperatures, certain bodies can conserve electricity indefinitely.

In 1908, the Dutch physicist Kamerlingh Onnes succeeded in liquefying helium, but at a temperature which had never before been reached: $-268.9°C$, that is, almost $4°$ above absolute zero. This technical achievement opened up the possibility of studying the properties of matter, and in particular metals, at very low temperature. It was also to allow the answering of a question in theoretical physics which was then being debated: was electrical conductivity maintained at very low temperatures, or did resistance decrease proportionally?

'Freezing' electrons

Onnes began his experiments with gold, then platinum, and observed that conductivity was just about maintained, which he attributed to the presence of impurities in his samples. Then he turned to mercury, a metal that can be purified to a very high degree by the technique of distillation. In 1911, Onnes was amazed to discover that at $-268°C$ the mercury lost its resistance almost entirely. The metal became superconductive. From 1911 to 1913 this most astonishing discovery was subjected to every possible effort to disprove it. Onnes carried out one particularly crucial counter-experiment: he introduced a current into a circuit kept at a very low temperature and then switched it off. Two years later there had been practically no loss of electricity. That meant, broadly speaking, that very low temperatures 'froze' the electrons

inside the matter. In 1913 Onnes received the Nobel prize for physics.

In 1933 the Germans Meissner and Ochsenfeld made a complementary discovery: at very low temperatures the magnetic field within a system in a state of superconductivity also becomes zero. Since there is electricity, even though confined, there ought also to be a magnetic field, but in superconductivity the currents tend to circulate on the surface of the superconductor and produce a field opposite to the external field. For this reason a superconductive body is said to behave at one and the same time as a perfect conductor and as a perfect diamagnetic body.

A third spectacular discovery stemmed from that of Onnes. In 1945 the Soviet Arkadiev tossed a magnet on to a sheet of superconductive lead: the magnet did not fall on to the sheet but floated above it in a state of weightlessness. That is explained by the induction of a secondary magnetic field, this one repulsive, in the sheet of lead, due to the creation of circular currents induced by the proximity of the magnet. The experiment took place, not at the surrounding temperature, but in a container filled with liquid helium.

'High' temperature superconductivity
Superconductivity for temperatures in excess of $-250°C$ was first observed in 1986 by Argentinian physicist Alex Müller and German physicist Georg Bednorz, using a ceramic of copper oxide containing barium and lanthanun. Other ceramics have been found to superconduct at temperatures greater than $-180°C$, but as yet no commercially useful high-temperature superconductor is available, because of low initial fields and currents, plus the brittle nature of the material.

Synthetics (First synthetic materials)

Braconnot, 1832; Pelouze, 1836; Schönbein, 1842. . . .

The plastics of long ago

The story of synthetic materials —
'plastics', as they are now called —
is much older than is sometimes
supposed.

The first synthetic materials appeared at the height of the Industrial Revolution, at the time when chemists were experimenting rather haphazardly with new combinations of products, guided sometimes by their knowledge, often by their instinct, and sometimes by sheer luck.

An explosive beginning

It was luck which helped the Frenchman Braconnot from Nancy, when in 1832 he poured concentrated nitric acid on to cotton fibres, then on to wood fibres. It should be made clear that, at the time, he had no idea what the result would be. Although known to alchemists like Albert le Grand since the 13th century (when it was prepared by heating nitre, that is salt-petre or potassium nitrate, with clay), nitric acid, the aqua fortis of the Ancients, was thought of only as a powerful solvent, hence its archaic Latin name *aqua dissolutiva*.

Dangerous billiard balls

What launched the American Hyatt into synthetic material research was a 10 000 dollar prize offered by the producers of ivory billiard balls to whoever could find a substitute material. Hyatt made balls from solid celluloid, covered with a layer of nitrocellulose. The result was that balls striking each other on the green baize exploded.

Perhaps Braconnot was expecting this acid (HNO_3) to dissolve the cellulose; nothing of the kind happened. In addition to the cellulose fibres the chemist obtained an impermeable film which he called xylonite. This was without doubt the first of all the synthetic materials. In fact it was one of the nitro-cellulose plastics, the basic constituent of numerous explosives, and, in 1836 another Frenchman, Pelouze, discovered the explosive properties of xylonite when it was heated. Exactly 40 years after Pelouze, the Swede Alfred Nobel was to register a patent on nitrocellulose by-products, focusing on dynamite.

Nobel might have been beaten by three decades at least for in 1842 the Swiss Schönbein, experimenting in his kitchen in England, spilled an acid, wiped it up with a cloth and put the cloth to dry on the stove. The cloth disappeared in a smokeless explosion. Schönbein repeated his experiment with the same results, then went straight to the Arsenal at Woolwich to demonstrate a new smokeless powder which would change the technique of manufacturing munitions. In fact Schönbein had discovered cellulose nitrate or guncotton, which was in fact to do well as an explosive. In 1846 Schönbein made, in the course of experiments and again by accident, the first transparent paper, cellophane.

Chance was working in favour of plastics; in 1838 the Frenchman Regnault, having left vinyl chloride in the sun, noted that it had resinified or, more exactly, polymerized, at least according to the definition of polymerization given in 1833 by the famous Swedish chemist Berzelius: the formation of a new substance made from two substances of different molecular weights, but in which the proportion of the same atoms has remained constant (see p. 160). In 1839 the Frenchman Simon perfected polystyrene and in 1843 the German Redtenbacher produced acrylic acid.

The drawback of these products is that they yellowed, twisted and decayed; in short, they were unstable. Several years were to go by before the American Baekeland discovered that the stability of polymers depended on the pressure at which polymerization occurred.

Collodion and Parkesine

Meanwhile, chance once again combined with inventive genius to produce a new synthetic material. Around 1845 the Englishman Alexander Parkes, an extraordinary dabbler in many areas — who, two years before, had patented the vulcanization of rubber by cooling (see p. 221) and the waterproofing of coats with rubber dissolved in carbon disulphide (an invention which he made over to the Macintosh company) — also became interested in nitrocellulose, but in the form of collodion, then widely used in photography. Collodion is actually nitrocellulose dissolved in alcohol and ether. It was used in support of silver bromide when it was applied to sheets of glass. Parkes imagined that thick layers of collodion would be rigid enough for the glass to be done away with, which was a forward-thinking idea; it is not known how he got from that point to finding that collodion could form a plastic substitute material, for example in making the artistic

mouldings for which he had a liking. Parkes spent many months trying to perfect this material and managed it, only without knowing it. The main drawback of his research consisted of the extreme inflammability — and capacity for explosion — of the basic materials, guncotton and the solvents which he used. Then he added camphor to his complex mixtures, by chance, and the camphor preserved the required plasticity in his material until it was cast in its final form. Parkes finally obtained a plastic which he could make into any shape and colour it as he wished. He called it Parkesine. It was very inflammable and its life was short, the factory producing it closing in 1870.

Legend has it that his successor in this field, the American Hyatt, accidentally rediscovered collodion as a possible support for synthetic material, when he found that a flask of collodion had overturned and formed a rubbery puddle (which is quite likely). What is certain is that Hyatt produced celluloid from nitrocellulose.

Composites, heirs of the synthetics

The term now used to denote new synthetic materials is 'composites'. These are made up of a matrix of resin, obtained from organic liquids which bind the support fibres in glass, graphite or in aramid organic materials. It is likely that an increasing number of cars, for example, will be made of composites, from the wheels to the drive shaft. The great advantage of composites over more traditional materials is that they are just as resistant, if not more so, but much less heavy and much more economical. Aviation is now beginning to take an interest in composites, which are also less vulnerable to fracture and corrosion.

Television (Origins of)

May, Willoughby Smith, 1873; Nipkow, 1880

It could have been invented much earlier

Selenium's sensitivity to light was the first fact leading to the invention of television.

In the 1870s, selenium, discovered a little more than half a century before (see p. 190), was widely used in the manufacture of current rectifiers, and so of measuring equipment for electricity. It was also used in telegraphic transmission equipment for the same reasons; that is, the transformation of alternating current into direct current. One day in 1873 an employee of a telegraphy station off the coast of Ireland, Joseph May, noted that the readings on his measuring equipment varied inexplicably according to the amount of sunshine. He discovered that the variations were proportional to the intensity of the sunshine.

Photoconductivity

In the same year the Englishman Willoughby Smith scientifically established that in fact selenium's conductivity is considerably increased by the effect of light and that this element immediately regains its resistance when the light source is cut off. It was the first known element to exhibit this effect, which was called photoconductivity. The design of current rectifiers was then amended and several types of photoconductive cells were launched. Shortly afterwards, in 1877, the Britons Day and Adams discovered that the same phenomenon occurred when selenium was brought into contact with a metal. They had established the photovoltaic effect, six years later

the American, Fritts, created the first photovoltaic cell, whose structure has remained fundamentally the same until now. This first cell worked satisfactorily but this is not really the reason why its principle has not been changed: the fact is that it was forgotten for nearly half a century.

The photovoltaic cell fired anew the imagination of the scientists, who had been dreaming since the 1840s — that is, since shortly after the American Morse had sent his first telegraphic message — of a way of transmitting pictures over a distance. It was then that, quite independently, the German Nipkow discovered for himself the effect of persistence of vision which makes an image persist on the retina for a longer period than the time of its actual transmission. He invented and patented a perforated disc whose very rapid rotation reconstituted an image by projection of points of light of different intensities. Nipkow had in fact invented the scanner. It was with a system derived from this, and improved by the cathode ray tube, that the first image was transmitted over distance. Television was on its way.

Photovoltaic effect
The production of electric current by light falling on some material, usually a semiconductor. The charge carriers are produced by photo-ionization. The effect is the basic mechanism of solar cells.

The first picture
The first picture transmitted over a distance was a portrait of the American President, Warren G. Harding, in 1923.

Transmutation

Rutherford, 1919

The alchemists' dream

The transformation of one body into another is possible through modification of its nucleus.

Transmutation, or the changing of one body into another, was an alchemist's dream, the origin of which seems to have been in China in the 3rd century BC. This project's basic aim was to transform base metals into precious ones. There is absolutely no reason to think that lead has ever been turned to gold.

The first scientific transmutation was accidental and occurred during a series of experiments into atomic phenomena. In 1907 the British scientist Ernest Rutherford had just demonstrated, with his compatriot J.T. Royds, that the particles of alpha radiation are made up of nuclei of helium, a significant discovery which earned him the Nobel prize in 1908.

The question then exercising the minds of the world's physicists concerned the structure of the atom, and the alpha, beta and gamma rays of radioactivity appeared to be some of the best tools with which to explore this structure.

Deviant rays

In 1911 the Englishman J.-J. Thomson put forward a theoretical model for the atom, in which the electrons divided up inside a positively charged sphere circulating in orbits around the nucleus. Rutherford then had the idea of using alpha rays to verify this hypothesis. According to his predictions, the alpha rays would have to undergo a deviation of 1–2° on crossing the atoms. The equipment thought out for the experiment was astonishingly simple: it consisted of a source of alpha rays perpendicular to an extremely thin sheet of metal. Two screens of zinc sulphide placed parallel to the metallic sheet — one in front of it, the other beyond it — were to allow for the picking up of the alpha particles reflected by the sheet, and for the picking up of those deflected after crossing the sheet. In the great majority of cases the deviation was of the order predicted by Rutherford, but in some others it reached 90°, which was extreme.

Intra-atomic forces

Rutherford himself explained these exceptional deviations by the fact that the alpha particle had not crossed the atom but had struck the nucleus, a nucleus small in size

---

Newton's illusion

Belief in the possibility of the transmutation of base metals into gold persisted until the 19th century. Newton devoted many years to it. It was not until the 1970s, however, that it became possible to obtain a few micrograms of an isotope of gold by bombarding mercury with rapid neutrons in such a way as to eliminate a proton of it — at great cost, as might be imagined. The intuition of the ancient alchemists was surprisingly accurate, for the two metals closest to gold in terms of atomic number are indeed mercury and lead. ... These would therefore be the most easily transmutable.

but with a mass almost equal to that of the whole atom and whose charge was identical to the sum of the charges of the electrons.

These experiments, made admirably clear by Rutherford, helped to establish a difference between heavy nuclei and light nuclei. Consequently, in fact, it was established that the impact of alpha particles on atoms with light nuclei brought into play a new type of force, intra-atomic forces, and it later turned out that these forces could modify the very structure of the atom.

Rutherford certainly was the first to have any inkling of this, for he continued his experiments of bombarding atoms with alpha rays. In 1919, at the Cavendish Laboratory in Cambridge, he bombarded nitrogen in this way and saw a few hydrogen ions or protons appear. For the first time he had succeeded, albeit unintentionally, in the transformation of one element into another: nitrogen into hydrogen, a phenomenon parallel to the natural formation of heavy water (see p. 88).

The first transmutation

The value of the discovery went far beyond the achievement of the alchemist's dream; it lay in the fact that the kinetic energy of the protons emitted by the nitrogen atoms could be greater than that of the alpha particles concerned. For this to be the case, the source of the energy difference would have to be the very nucleus of the nitrogen atom.

The actual process of transmutation obviously intrigued the physicists, who repeated Rutherford's experiment in many different ways. In 1923 the Englishman

---

Becquerel, the forerunner

A share of the credit for the discovery must go to the Frenchman Henri Becquerel, who first observed that the radioactive substance uranium, which he called uranium X, lost its activity over time, implying a specific modification. Rutherford's merit and that of his compatriot Frederick Soddy lies in having formulated the theory of subatomic chemical transformation, after having observed the same phenomenon in thorium.

P.M.S. Blackett, observing some 400 000 trajectories of alpha particles on nitrogen atoms, recorded eight cases of transmutation. Blackett had just furthered Rutherford's experiment in a fundamental manner, for the reaction that he had noted:

$$^{14}_{7}N + ^{4}_{2}He \rightarrow ^{17}_{8}O + ^{1}_{1}H$$

not only gave hydrogen, but also oxygen. There had been transmutation of one element into two others. Such a result could only be explained by a reorganization of the atoms themselves.

Fission

In 1919, Rutherford had touched on the phenomenon of fission, discovered with disbelief 20 years later by Otto Hahn (see p. 23), on the strength of the reorganization of energy in the bombarded atom, but he had not grasped the full meaning of it. It must be said that Rutherford, like Blackett and later Irène Curie and Frédéric Joliot, were the first, very distant, pioneers of the idea of a release of energy from the atom and that they were particularly concerned with understanding the energy changes which occurred in the atom under the influence of ionizing rays. The work of Curie and Joliot on artificial radioactivity (see p. 20) was to bring Rutherford's discovery to fruition. In 1934 the two scientists produced isotopes of phosphorus and silicon by bombardment of aluminium according to the reaction which caused a sensation in the scientific world:

$$^{27}_{13}Al + ^{4}_{2}He \rightarrow ^{30}_{15}P + ^{1}_{0}n$$

the phosphorus with a mass of 30 emitting $\beta+$ radioactive radiation of about 3 minutes, to produce a second transmutation:

$$^{30}_{15}P \rightarrow ^{30}_{14}Si + ^{0}_{1}e +$$

the unstable phosphorus having produced, by loss of a positive electron or positron, a stable isotopic silicon.

The lessons learnt from this brought the physicists closer to fission, since it showed that alpha bombardment of a body could end in the production of the more powerful beta radiation, with emission of neutrons and positrons.

Most of the scholars who took up Curie and Joliot's experiments, the Americans Ellis and Anderson, and the Germans Meitner and Frisch saw in them above all the possibility of creating new radioactive elements.

The ideal projectile

At that point the fundamental discovery of fission was only five years away, although no-one had as yet the least inkling of it. In 1932 the Britons Cockcroft and Walton built a particle accelerator which helped to increase very markedly the energy of natural ionizing particles by a series of successive accelerations: having bombarded lithium with ultra-rapid protons, they achieved its transmutation into unstable beryllium, which itself divided to produce two nuclei of helium, that is to say alpha rays. That same year, 1932, the Englishman Chadwick made a fundamental discovery, which was that of the neutron, a particle with a mass almost equivalent to that of the proton but possessing no charge; this made it an ideal atomic projectile, since it was capable of penetrating the atom without being affected by either the charge of the nucleus or that of the electrons, that is to say, indifferent to the charge balance of the atom and capable of splitting the nucleus.

Still in the hope of creating new radioactive elements, numerous physicists, including Curie, Joliot, Fermi, Rasetti and d'Agostino, had in fact begun to bombard uranium and many other elements with neutrons, and they obtained apparently new elements with atomic numbers beyond uranium; these elements seemed very strange.

The new alchemists

It had been well known, since the work of Fermi, that the neutron projectile was captured by the irradiated nucleus and that the new nucleus formed produced a powerful gamma radiation; this placed it beyond uranium in Mendeleyev's periodic classification, but no-one really suspected what was actually produced. That was until the day in 1939 when Hahn achieved the 'transmutation which transmuted transmutations', so to speak: bombarding uranium with neutrons, he obtained barium and argon! The dream of the alchemists had been vastly surpassed: it turned out that what had been happening since Rutherford, without anyone realizing it, was the splitting of the nuclei of bombarded elements. The dream of the Spanish philosopher Raymond Lull had finally resulted in the reality of atomic energy.

Troy

Schliemann, 1872–4

The birth of archaeology

*The discovery of the legendary city
taught archaeologists the historical
value of legends.*

The discovery of Troy, capital of the kingdom immortalized by Homer and over reigned by the unfortunate Priam, was a turning point in the history of archaeology. It comprised both a negative and positive aspect, which have subsequently provided fruitful lessons for archaeologists.

The negative aspect was due to the personality of the discoverer, the German Heinrich Schliemann, the most celebrated of all archaeologists. A rich merchant born in 1822 in the Mecklenburg-Schwerin district, fed since childhood on the splendours of the Homeric legend, Schliemann threw himself into an assault on the site at Hissarlik in Turkey; he was inspired by a passion which was hardly scientific and which indeed had elements of megalomania, naivety and greed, all coloured with a certain disregard for the truth. He considered that the remains of Troy, which he was confident of finding, belonged exclusively to him, for he had made it his dream. An archaeologist of no real training, his work on the ground was rather chaotic: he crossed stratigraphical levels (vertical sections through the earth showing the chronology of occupation) with frenzied determination, decided, for example, that

the 'Mycenaean' fortress, the remains of which were found at the second level, was the ancient Pergyma, that the gold and silver jewels, the statues, weapons and vases — which, it is maintained, were gathered in different areas of the site and not in a single tomb — constituted the 'treasure of Priam', saw traces of Hector here, traces of Paris there, and finally appropriated the treasure which he took out of Turkey, an outrage against the Sublime Porte (Ottoman government). His discovery was the unexpected fruit of a visionary mind and a good knowledge of history, but unfortunately it constituted a kind of 'crude' archaeology which is little more than pillage.

An inspired innocence

If it were not for the much more professional work of his old collaborator, the architect Dörpfeld, in 1893–4, and also the much more scientific contribution of the American Blegen from 1932–8, not much would be known about Troy. The lesson to be learnt here is that every excavation must be carried out according to scientific procedure, which takes precedence over any private enthusiasms.

Schliemann was not, however, entirely lacking in method, at least in his identification of the site. He carefully studied the 'historical' evidence (in fact it was literary evidence, for these were the reports of chroniclers, since history in the modern sense was not yet born), as well as Homer's

---

A passion for Homer

Schliemann was so passionately interested in Homeric legends that, married to a Greek who was also his collaborator, in fact his deputy, at Mycenae, he called his children Andromache and Agamemnon.

---

text, the information from Herodotus, Thucydides and Strabo, an analytical geographer who compiled a register of ancient data, from Demetrios of Scepsis and many others. He then concluded that the citadel of Troy was actually to be found below the hill of Hissarlik in Asia Minor, on the coast opposite the Dardanelles, on the promontory between the river Scamandre (now the Menderes) and its tributary the Simois (now the Dumbrek Su). As early as the year 160 controversy over the real site of Troy arose with Demetrios of Scepsis, but perhaps this is because the site has been modified by earth tremors of long ago. The Scamandre, for example, has changed its course.

Whatever the cause, Schliemann, steeped as much in culture as in fantasy, was gripped by a 'logic of locations'; he recognized to the left the legendary peak of Samothrace, to the right, Mount Ida. By instinct he reconstructed the movements of the Acheaen troops from Greece, who disembarked a short distance away. When Schliemann dug his first spit in 1871, so launching the first of his four campaigns, Troy had, as far as historians were concerned, dissolved into the mist of legend. After 1872 it was clear that he had found a very ancient site, even if it was not until 1874 that he laid his hands on the treasure with which he was to make off.

The scandal caused by this plundering of the spoils — for which Schliemann was to compensate the Turkish government amply — forced him to abandon the excavations. Fortunately Dörpfeld was able to find sufficient funds to take them up again; he found the great citadel close to the small acropolis excavated by his predecessor, which indeed seemed to be a royal palace.

Enriched by the treasure plundered from Troy, Schliemann next turned his attention to the site of Mycenae in Greece, again basing himself on ancient writings, this time those of Pausanias, according to which Agamemnon, conqueror of the Trojans, was buried there. In this desolate place he dis-covered the most fabulous treasure ever brought to light in archaeology; if it were not for him the Mycenaean civilization would have long kept its secrets.

History and legends

Like the discovery of Troy, that of Mycenae included one very important lesson for modern archaeologists: this was that ancient texts, neglected for a long time and considered as collections of dubious or 'literary' legends, may contain information of historical value. Formerly held to be simply an epic masterpiece, the *Iliad* is now studied, along with many other texts, as a source of serious historic information, despite its innumerable references to divine interventions.

Armed with knowledge drawn from similar texts, notably from the Bible, very many archaeologists have undertaken digs and studies of sites and events formerly considered as legendary or questionable. So, since the 1950s, archaeologists, especially American ones, have systematically explored the Arabian peninsula in search of pre-Islamic kingdoms, such as that of the Queen of Sheba, the legendary admirer of Solomon whose existence has been restored to the realm of great probability if not of certainty.

Schliemann's treasures

Housed in the Berlin Museum, the treasures of Troy 'recovered' by Schliemann were thought to have been destroyed by the bombings of 1944. It is reported that they are in the USSR. The treasures of Mycenae, however, jealously guarded as they emerged and deposited in the Bank of Athens, are now in the National Museum of that city. Despite the scandal caused by his hijacking of the 'treasure of Priam', Schliemann was allowed to return on several occasions to Turkey and to Hissarlik.

Tungsten steel

Köller, 1855

The first modern alloy

Nearly half a century separates the first successful ferro-tungsten alloy from the earliest production of objects made from it.

Towards the middle of the 19th century scientists and metallurgists were intrigued by an ore which, because of its great density, was very heavy. It had been discovered by the Swede Scheele, and so was called Scheelite, but people also continued to call it by the name created by the Dane, Cronstedt, from the words *tung*, which means heavy, and *sten*, stone: tungsten. In 1847 the metal was successfully isolated but no-one had any idea what to do with it.

An unexploitable asset

At that time the rapidly expanding iron and steel industry was already seeking to produce high-performance steels, but it was still impossible to manufacture more than a few dozen kilograms of iron or steel at a time and conditions were extremely

---

The most expensive bullets in the world

The most costly application of Köller's discovery was in the military field. During World War II Germany manufactured bullets from tungsten steel. These had the distinction of being able to pierce armour plating of 2–3 cm but also had the drawback of being very expensive. They were abandoned, only to be taken up later by the United States and France. These bullets are currently produced for use by the French army and are priced from £1–£15 each, according to whether they are intended for use against light or heavy vehicles.

---

unpleasant. In 1855 the Austrian Köller had the idea of incorporating tungsten into steel during smelting. This was the first alloy of its kind, and produced steel which was very resistant but very difficult to work. The following year, after the introduction of the Bessemer converter and the Siemens furnace, it was found that the addition of manganese produced steel which was both very resistant and easy to work. It would have seemed that there was no further use for Köller's discovery, but it was made again in 1857 by the Englishman Robert Mushet and he took out a patent on his discovery. It remained unexploited for a long time.

In fact the industry used alloys of manganese, nickel and chrome for several years before turning again to the possibility of a tungsten alloy. It was only at the Paris Exposition in 1900 that the Bethlehem Steel Corporation of the United States unveiled the first tools made in tungsten steel.

It was much later and thanks to the introduction of electric furnaces that the potential of tungsten steel was fully realized. The favourite use for this alloy was to be in cutting steels. These might be composite steels with a base of tungsten, molybdenum, chrome, vanadium or niobium, or ferrotungstens with 78 per cent tungsten or cemented carbides.

The growth of the car industry, which was already demanding steels capable of withstanding high temperatures and mechanical pressures without becoming distorted or breaking, contributed greatly to the development of tungsten steels. The reason for the delay between the discovery of tungsten steel and its industrial application is bound up with the need for temperatures in the order of 1400°C and for complex equipment in the manufacture of the alloy. Tungsten steel, the first of the ferrous alloys to be discovered, was therefore the last to be fully developed.

Uranus

Herschel, 1781

Suspected before being seen

This was the first planet to be discovered since antiquity.

The most ancient astrological texts mention five planets: Jupiter, Saturn, Venus, Mars and Mercury. Astronomical telescopes (see p. 22) improved sufficiently, at least in principle, from the 18th century on for there to be no reason why others should escape observation.

Perfectionism

The discovery of Uranus did not stem from deliberate research, as was the case with Pluto; it was rather from the general curiosity of an 18th-century astronomer, the Englishman William Herschel. A musician of German origin living in Bath, Herschel was initially, like his sister Caroline and his brother Alexander, an amateur astronomer; but he was an amateur of the first rank, who took perfectionism to the point where he ground the mirrors of his telescopes himself and jealously polished them with great care. Herschel made instruments of exceptional quality, superior even to those of the Observatory at Greenwich. Therein lies the secret of his discovery.

At the age of 43, Herschel decided to become a professional astronomer and set about writing a 'natural history' of the skies. This gave him a particular interest in nebulae, or at least in those objects which his colleagues called nebulae. Armed with his best telescope, which had a magnification power of 6450, he found, on 13 March 1781, a 'curious nebulous star or perhaps a comet', which he quickly realized was not, in fact, a star because its disc shape was sharply outlined. Neither was it a comet, for it had no tail. Its movement strongly suggested that it might be a planet.

Herschel observed it for a year and discovered that its orbit was indeed planetary and that its radius was 18 astronomical units (one of these units being the mean distance from the Earth to the Sun).

The excitement following the discovery of this unknown planet in the solar system was immense, and Herschel became famous. The intrinsic interest of the discovery was unquestionable. Uranus, a variable source (its brilliance varies by 15 per cent in 10 hours), continues to arouse the interest of astronomers. It was also significant in that it revealed, through the anomalies of its orbit, the existence of its 'quasi-sister' Neptune, discovered in 1846 (see p. 136).

Pursuing his observations, Herschel discovered in 1787 two of Uranus' satellites, Oberon and Titania, the first two of the 13 satellites discovered up until 1986 (the last eight having been discovered by the American space craft *Voyager II*). They have been given names in the Shakespearean tradition, as begun by Herschel.

Georgienne or Uranus

Herschel at first proposed to call this planet 'Georgienne', in honour of King George III, but in the end the idea of naming it Uranus, the father of Saturn and grandfather of Jupiter in Roman mythology, prevailed.

Not really a discovery

Although Herschel must be considered to be the discoverer of Uranus, being the first to identify it as a planet, the celestial object itself had already been observed 17 times before, the first time in 1690. The first observer to note the planet had been the English astronomer John Flamsteed. That and the subsequent 16 observations had led astronomers to the conclusion that this brilliant object was a fixed star.

Vaccination

Anon, c. 10th century; Montagu, 1718; Pasteur, 1881

Using the bad to good purpose

Protection against infectious diseases by inoculation with the germs themselves goes back to antiquity.

Derived from the Latin *vaccinae*, literally 'of the cow', the word vaccination appeared around 1880; it seems to have been used for the first time by Pasteur on the basis of work done by the Englishman Jenner, who had immunized patients against cowpox and common smallpox.

Vaccination is not, properly speaking, a discovery, for it had been used in Turkey since time immemorial. It is possible that it resulted from the experiments of the first known toxicologist in history, Mithridatus IV, King of Pontus (north of modern-day Turkey), who claimed that one could be immunized against poisons by regularly absorbing small quantities of them. The Turks 'vaccinated' against smallpox by extracting traces of the contents of the pustule from mild cases of the disease and injecting them into healthy people.

This rather risky practice was discovered by the wife of the British Ambassador to Constantinople, Lady Mary Wortley Montagu, a leading light in international society at that time. Lady Montagu introduced this type of immunization, called 'variolation' (from variola, another name for smallpox) into Britain in 1718. Many people were immunized in this way, but some died from the effects of the vaccination. It cannot truthfully be said, therefore, that Jenner, who was the first to practise vaccination against smallpox on a large scale, from 1798 onwards, really discovered the principle of this therapy.

Smallpox

His discovery was based, moreover, on confused and even false principles. A country doctor, Jenner believed that cowpox and no doubt smallpox, similar but not identical viral diseases, had their origins in an infection of horses' hooves, from where they spread to cattle and perhaps to humans. The reality was that the disease was transmitted by infected farmers, who in turn infected the cows, in which the disease became cowpox.

The type of vaccination practised by Jenner was basically not much different from that which Lady Montagu had introduced, since at first the people who had been treated had to return to Jenner a week

---

Jenner's worth

His worth lay in having discovered that inoculation with the cowpox vaccine could immunize against smallpox.

Smallpox vaccination in China

Under the Chinese Song dynasty between the 10th and 12th centuries vaccination against smallpox had been practised, but it seems to have been abandoned by the time the Turks were using it and Lady Montagu imported it into Europe.

later so that extracts could be taken from the pustule for the inoculation of others and so on. However, this method gave rise to a problem, namely that the viral stock gradually weakened and several times Jenner had to inject human extracts into cows to revive the stock. This is what is called 'retro-vaccination'.

However, there was indeed one difference between the crude variolation introduced by Lady Montagu and that practised by Jenner: Jenner did not inject the smallpox virus itself but that of the vaccinia (cowpox) virus, which is different, but yet triggers effective immune reactions against smallpox. This was proved when Pearson, an ill-advised emulator of Jenner, also became involved in vaccination against smallpox and caused serious cases which were very similar to the disease he wished to prevent. In 1799, the incident having been used as ammunition by the opponents of Jenner's vaccination programme, Jenner demonstrated that the preparation used by Pearson had been contaminated by smallpox germs. Furthermore Jenner did not take his samples until the seventh day after the appearance of the pustules, that is to say, after the germ had lost its virulence. So it can be said that Jenner discovered the principle of vaccination by attenuated germs.

Jenner and Pasteur

In spite of a tide of hostility, Jennerian vaccination gained ground; in 1800 it was introduced into France by the Duc de la Rochefoucauld-Liancourt. In 1803 the Royal Jennerian Society was created in Great Britain, guaranteeing free vaccination to the public.

Meanwhile, the notion of attenuated germs made progress in the minds of doctors. It was obvious that one could not inject the actual germs of a disease which one wanted to immunize against, for fear of triggering the disease itself. Although almost all the mechanisms were then

unknown, the notion of immunity was beginning to make headway and it was claimed, justifiably, that inoculation with an attenuated germ could help the organism to recognize a germ and defend itself against it. This is why, when Pasteur prepared the first anti-bacterial vaccine against anthrax, he used attenuated germs. On that occasion the French scholar paid a warm tribute to Jenner as 'one of the greatest Englishmen'.

This principle of attentuation was implemented by Pasteur in his preparation of a vaccine against rabies: the virus used was the object of 100 successive intra-cerebral injections with a sample of material from the spinal cord of an infected rabbit, and from one animal to another. The vaccine had only been tested on dogs when, in 1885, Pasteur was presented with a 9-year-old child, Joseph Meister, who had been bitten by a rabid dog. Although not a doctor, Pasteur responded to the challenge: he used the vaccine on the child with successful results. Modern vaccination had been born.

The only important modification made subsequently was the introduction of vaccines obtained through genetic engineering. These were brought into use in 1983, the first one commercially produced being the vaccine against hepatitis B in 1986.

Jenner's dedication

Twelve thousand people were vaccinated in London during the 18 months following the opening of the Royal Jennerian Society. Jenner himself carried out some 300 free vaccinations a day. The average annual number of deaths from smallpox fell in the following year from 2018 to 622.

The end of smallpox

Smallpox seems to have been eradicated, with no cases since 1978, but it is suspected that some animal reservoirs might survive and unleash new epidemics.

Variable stars

Fabricius, 1596

A permanent discovery

It has been known for four centuries that certain stars do not have fixed brightnesses. The reasons for this are as variable as the stars themselves.

In 1596 the Dutch astronomer David Fabricius, son of the astronomer who discovered the first sun spots (see p. 196), observing the sky closely and regularly, discovered a star whose magnitude, or brightness, was not constant; this was Mira Ceti. The discovery was regarded as a curiosity but, cosmology and astrophysics being almost non-existent, it was of no consequence, apart from encouraging certain astronomers to check whether other variable stars existed.

So in 1667 the Italian Montanari discovered a second, βPersei or Algol. Then, thanks to an improvement in telescopes, more and more variable stars were observed: by 1885, 113 were known; by 1986 more than 20 000 had been counted. Fabricius' observation had opened a new chapter in astronomy.

The significance of the Cepheids

In 1912 the Englishwoman H.S. Leavitt discovered the existence of a relationship between the period of variability and the luminosity of a particular type of variable star, the Cepheids, which are characterized by variation over very short periods ranging from a few hours to a few weeks. This had very important repercussions for the determining of distances within galaxies.

In 1958 Kukarkin and Parengo proposed a classification of this type of star: it consists of three large groups: the true pulsating stars, divided into 22 types; the eruptive variables, divided into 12 types; and the eclipsing binaries, divided into 5 types. This classification reflects the basic differences in variable stars, including double stars, whose variation in magnitude is due to the periodic occultation of one by the other; novae (or eruptive stars); and those which could be called the 'true' variables, whose pulsations are due to internal phenomena.

The brightness changes of variable stars with a long period of variation can be in the proportion of 1 to 100, doubtless the explanation for differences in brightness observed by Fabricius in Mira Ceti. The variations in brightness can be at regular or irregular intervals. Stars with irregular pulsations are also called stars with outbursts or flares. These become several times more brilliant in a few seconds. The phenomenon presents numerous analogies with solar eruptions.

Viruses

Ivanovski, 1892

Stray bits of DNA

For nearly a century the way in which these parasitic germs act and the scope of their behaviour have continued to modify our notions about disease.

At the end of the 19th century the perfecting of bacterial cultures, achieved by Pasteur and Koch, gave rise to the hope that it might be possible, in the not-too-distant future, to study almost the whole range of germs with the expectation of finding treatments for them. Then, in 1892, a discovery by the Russian Ivanovski showed that the task was more complex than had been supposed: filtrates of juice from tobacco plants infected by a disease called mosaic were passed through a porcelain filter which Pasteur and Chamberland had just perfected. He found germs smaller than bacteria, therefore impossible to examine under a microscope.

Filterable germs

Ivanovski had noted that the filtrate had infected healthy tobacco plants. The new category of filterable germs, however, was not limited to the plant kingdom: in 1898 the Germans Löffler and Frosch in turn discovered that foot and mouth disease in cattle belonged to the diseases transmitted by this type of filterable particle. In 1901, the American Reed showed that yellow fever is also due to a virus, the term stemming from the Latin word meaning 'poison'.

Vaccination against viral diseases such as smallpox (achieved by Jenner; see p. 209)

and rabies, perfected by Pasteur, had become established practice but it was not yet known exactly what viruses were. For nearly half a century research would be limited to establishing a catalogue of viral diseases and, if possible, producing vaccines against them, but the field of investigation barely went beyond that of infectious pathology (the process of disease). In 1913, the German Grüter identified the herpes virus, which can trigger viral septicaemia or meningo-encephalitis in the newborn. In 1927 his compatriot Degkwitz did the same with the measles virus. The catalogue of viral

The factors of mutations

We do not know all the reasons which make a dormant, sometimes harmless, virus become 'virulent', but we have known since Luria in 1952 that ionizing and ultraviolet rays can provoke viral mutations. Some researchers, like the Russian Tchijevsky, established relationships between viral and bacterial epidemics and terrestrial electromagnetic variations such as can be caused by sunspots (see p. 196). The year of the terrible 'Spanish flu' epidemic, 1918, did in fact coincide with a peak in solar activity. Other factors, however, can also play a part: for example, the widespread production of antibodies to a type of 'flu virus can spark off a recurrence of the virus in a different form the following year.

infections increased as the years went by, to include German measles, chickenpox, 'flu, polio, mumps and hepatitis, as well as diseases which had been mysterious until then, such as Saint Louis' encephalitis, tonsillitis with glandular fever, certain lung diseases (with cytomegalovirus), haemorrhagic fevers like green monkey disease or Lassa fever, and so on.

Until 1934, virology was considerably hampered by the difficulty of studying viruses. They are very difficult to see using a microscope — the dimensions of even the largest ones, like the psittacosis virus, are in the order of 300nm, 1 nanometer equalling 1 millionth of a millimetre — and it is impossible to cultivate them. According to all the evidence a virus dies with the cell which it is infecting, although it was reported in 1986 that some viruses can survive for a few days in dead cells. That same year, however, the Englishman Elford perfected the technique of filters with graduated pores, which only lets particles of known diameter pass through. Also in that year, the American Goodpasture discovered that it is possible to cultivate certain viruses on chicken embryos. This discovery opened the way for his compatriot Enders to cultivate the polio virus, which led to the production of the first anti-polio vaccine (see p. 209).

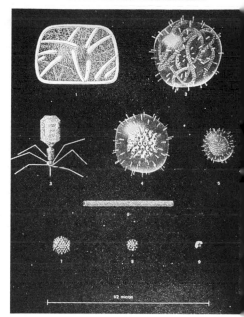

Different viruses

These new types of virus can only hint at the variety of forms and dimensions of these specific germs which, unlike bacteria, only survive if they manage to insert their genetic material into that of a living cell.

Viruses and DNA

The diversity among viruses, which proved to be extensive, prevented any decisive advance. All that was done was to classify them in four groups: those which are responsible for human and animal diseases; those which affect insects; those which are responsible for plant diseases; and the bacteriophages. This last group was established following the discoveries by the Englishman Twort and the Frenchman d'Hérelle between 1915 and 1917; they had revealed that certain viruses, termed bacteriophages, can infect bacteria. The main progress was recorded in the therapeutic field, and was based on the perfecting of new vaccines and on serological detection (that is, the detection of viruses, and bacteria, by their effects on the immune system). Developments in electron microscopy, which confirmed the great diversity of form and structure of viruses, also revealed that, although viruses are most often spherical or ovoid, their structures can vary considerably.

'Therapeutic' viruses

The type of virus called the bacteriophage has inspired numerous scholars to research into a method of antibacterial action, and even into the possibilities of genetic intervention, aimed at correcting hereditary faults like the blood disorder thalassaemia. These researchers are very cautious, for it is impossible to be absolutely sure in advance what the possible reactions of a virus will be to new circumstances, and a 'useful' virus may one day turn out to be harmful, following a mutation for example.

Major advances were made in the 1960s, thanks to disciplines of genetics and pathology. In the field of genetics they were the result of meticulous research by numerous international teams and it is hardly fair to single out one name to link to the major discovery of the 1960s: that the virus reproduces itself by injecting its genetic material into the DNA of the host cell, either directly if it is a virus with DNA, or indirectly if it is a virus with RNA, after production of a DNA. The DNA of the cell then reproduces the virus.

The second series of advances is due more specifically to discoveries in infectious pathology. In 1957 the Englishman Burkitt discovered that a cancer of the nose and throat, which is particularly widespread in Africa and China, (Burkitt's lymphoma) is caused by a virus of the Epstein–Barr group, to which the two herpes viruses belong, along with that of a very rare African cancer, Kaposi's sarcoma, and that of infectious mononucleosis or glandular fever. The information caused problems in all areas of medicine, among oncologists (cancer specialists), virologists and biologists, for the almost universal conviction of the time was that cancers are not caused by viruses. The case of Burkitt's lymphoma was classed among the exceptions which prove the rule, a rule based moreover on an extremely fragile theory, which was that no virus had ever been found in cancerous tissues. The truth is that no virus had been found because it was inserted into the cellular DNA, and because the methods of investigation were unreliable.

The idea of a viral origin for certain cancers is not new, however. It goes back to 1910, the year the American Rous discovered that *sarcoma* (a type of cancer) in the hen is due to a virus and succeeded in transmitting this and two other types of cancer to the animal by inoculation with a filtrate.

The viral hypothesis

When the Frenchman Borrel subsequently defended, on theoretical grounds, the hypothesis of a possible viral origin of cancer he met nothing but scepticism. Yet Rous' discovery was not unique: two years

before, in 1908, the Germans Ellermann and Bang, followed by the Frenchmen Oberling and Guérin, proved that avian cancers, like the leukaemias, and also some mammalian cancers can be transmitted by inoculation with filtrate. A distinction among oncologists between animals, specifically mammals, and man. Yet, until Burkitt's discovery, the hypothesis of a viral origin for certain cancers was discredited and research into this phenomenon was rare, if not non-existent.

Burkitt's discovery did, however, arouse suspicions and in the 1970s several American researchers noted a coincidence between female infections with genital herpes and cancer of the cervix, or neck of the womb. Interest in the viral hypotheses was rekindled, but soon died down in the absence of any new discoveries or research.

AIDS

However, in 1981 interest was suddenly reawakened by the appearance of a new and alarming disease, the acquired immunodeficiency syndrome, or AIDS. An extraordinary fact is that AIDS shows up in many cases with the appearance of Kaposi's sarcoma; it was in fact the abnormal frequency of acute cases of this cancer, which until then had only been observed in West Africa and only in slowly evolving forms, which alerted the Center for Disease Control in Atlanta, an extremely well-equipped epidemiological unit, to its presence. All the characteristics of this multiform disease, whose apparent common denominator is paralysis of the immune defences by destruction of the T4 lymphocytes (a type of white blood cell), suggest that it is of a contagious nature. In 1984, in the space of three months, the Frenchman Montagnier of the Pasteur Institute and the American Gallo identified the agent responsible for AIDS: it is the human immunodeficiency virus (HIV), a virus which is declared by Montagnier to belong to the LAV family and by Gallo to belong to that of the HTLV family. Each man based his theory on his personal interpretation of the structure of this virus and on this point it must be noted, in support of Montagnier, that, unlike the

HTLVs, it does not cause an increase but, on the contrary, a decrease, in the levels of T4. This is the second piece of evidence proving viruses responsibile for certain cancers. (In the case of Kaposi's sarcoma, it seems that the virus acts as a co-factor of the Epstein–Barr virus.)

Studies completed in 1985, again by Montagnier, Gallo and many others, revealed two other facts about the AIDs virus. In the first place it is a slow virus, belonging to the same family as those which cause Creuzfeldt–Jacob disease (an irreversible degeneration of the brain leading to early dementia) and an exotic disease, kuru, which has now disappeared but was associated until the end of the 1950s with cannibalism in Papua New Guinea. Secondly — and this information was of considerable surprise to both virologists and oncologists — it is a retrovirus, that is to say an RNA virus. Now this second fact was a considerable breakthrough: until then all the virologists had supposed that retroviruses did not infect human beings. Proof came in 1985, however, of the possibility of humans being infected with the AIDS virus.

Virology and genetics

In 1985–6 a theory gradually took shape: it was that the AIDS virus might be a mutant form of that responsible for an red monkey disease, dramatically discovered in the 1960s. The mutations themselves had been recognized since the discoveries of the American Luria in 1952, made on bacteriophage viruses and done by others on viruses such as that of influenza. A virus can change character; the influenza virus, for example, possesses several potential genetic forms and it is accepted that the epidemic of 'Spanish 'flu' of 1918, which caused some 20 million deaths in a few months and affected nearly half the planet (in India it reduced the level of the population by 4 per cent in five months), could have been caused by a mutant form of the swine 'flu virus. In 1986 it was finally

acknowledged that half of all cancers could be due to viral actions.

The discovery of prion in 1982 (see p. 162) and numerous other work in progress gave rise to the idea, in 1986, that the problem of viruses touches on questions of fundamental genetics. If it is now accepted that definite populations of virus exist which are very close to bacteria, numerous geneticists admit that other viruses might represent intermediate forms of 'incomplete' life; they would, as it were, be fragments of DNA overlooked in the great genetic distribution of living species and trying to fit themselves in rather belatedly. The existence in the living being of non-infectious proviruses, which only become infectious in conditions which are not yet established, through genetic recombinations, with the aim of organizing the DNA of the host cell to their advantage, indeed the possibility that unexpressed fragments of the human genetic capital might trigger the formation of viruses, and other indicators show that virology is closely identified with the study of the mechanisms of life itself. The discovery made by Ivanovski in 1892 hardly paved the way to the many great triumphs of contemporary research.

The mysteries of virulence

If mice are injected with massive doses of one or other of two forms of the non-virulent viruses of herpes, a viral infection, they are not affected by it. If they are injected with the two forms together, though, 60 per cent of the animals die, as was shown by the experiments of a team of virologists from the University of Los Angeles in 1986. This phenomenon is probably explained by the fact that the two strains which are non-virulent in isolation become virulent when in contact with one another. It is supposed that the key to all this lies in a genetic recombination. It might partially explain the sudden virulence of certain germs and the appearance of new viral diseases like AIDS.

Viscose

Audemars, 1855

Artificial silk

---

*The first synthetic material was the
product of chance.*

Viscose, the first synthetic material based entirely on cellulose, was the product of a very old industrial dream and of efforts of successive, more or less fortunate researchers. The first to foresee the possibility of manufacturing synthetic materials was the Englishman Hooke who, in his *Micrographia*, published in 1664 and therefore well before the birth of organic chemistry, raised the possibility of copying the secretion of cellulose threads by the silk worm. The discovery of gums then led the Frenchman Réaumur, in 1734, to think of artificial cellulose fibres.

The first to actually produce anything was the British weaver Schwabe who, in 1842, poured molten glass through a fine sieve and obtained filaments which he wove. The result was not very satisfactory but Schwabe had found the principle of production for what were to be the cellulose fibres of the future. In 1846 the German Schönbein had the idea of applying Schwabe's principle to the production of threads drawn from nitrocellulose, which he had discovered himself. The textile obtained from this was still not satisfactory.

In 1855, the Frenchman Audemars nitrified some wood from the mulberry tree, the tree associated with silk worms, and then mixed it with ether, alcohol and rubber solution. This unexpected mixture was not

sieved according to Schwabe's principle, but transformed into threads by an original method: the point of a needle was dipped into the mixture and the hardening thread which was attached to the needle was woven! The resulting textile must have demonstrated some useful qualities, for Audemars obtained a patent for the production of his unusual fibre.

Towards artificial silk

Two years later, trying almost anything, the Englishman Hughes mixed fats, resins, tannins, gelatine and starch and obtained a fibre which at the time was claimed to resemble silk. . . .

The next attempt was hardly more conclusive. As cellulose nitrate was highly inflammable because of the combination of its two elements, and as the process of denitrification by aluminium sulphide had been perfected in 1863, the Englishman Swan hit upon the idea of producing a fibre of denitrified nitrocellulose, the process of nitrification having transformed the cellulose into a paste from which the threads

---

Rayon

When, in 1924, viscose was renamed rayon, the artificial silk industry, relying on the four processes mentioned, was at its peak. From socks to electric wire coverings, from carpets to car covers, from surgical materials to bed linen, it established itself as an entirely separate industry and the term 'artificial silk' fell into disuse.

Saponification

As used by Miles, the hydrolosis (decomposition of organic compounds by interaction with water) of esters into acids and alcohols by the action of an alkali (caustic soda).

could be drawn. The idea did not catch on.

In 1878 the Frenchman Chardonnet took up the research and discovered that one of the greatest difficulties in the production of dry nitrocellulose fibres — the coagulation of the paste filament — could be overcome very easily; in fact, working his mixture over heat, he noted that this in itself was enough to dry the filament. In 1884 he undertook the industrial production of woven fibres of what he called viscose, preparing his paste from mulberry wood. At the Universal Exhibition of 1889, where Chardonnet displayed his first viscose textiles, the material was called artificial silk. It had considerable disadvantages: it was highly inflammable and, once wet, it lost its shape.

The stretching of filaments

Then events accelerated. In 1890, one year before Chardonnet set up his viscose factory in Besançon and by dint of trying one process after the other, the Frenchman Despeissis independently rediscovered a technique which had already been discovered three times: cuprammonium hydroxide, an ammoniacal solution of copper oxide, dissolves cellulose as well as nitrate. (The earlier discoverers were the Englishman Mercer around 1850, the German Schweitzer in 1857, and the Englishman Weston in 1882.) The Germans Pauly, Fremery, Bronnert and Urban perfected the process, which would be used industrially by the Bemberg company from 1898, but would not become profitable until after 1919, with the introduction of the process of stretching the filaments. At the same time the British researchers Cross and Bevan, who had been on the track of cellulose acetate since 1884, discovered in 1892 that cellulose can be manufactured in fine textiles by another method: some cellulose xanthate is put in solution into simple caustic soda. Unfortunately, the filament obtained was neither stable nor easily woven, and it was not until 1898 that the Englishman Topham discovered a means of stabilizing it. In 1902 he invented equipment which produced usable threads of cellulose: it consisted of a perforated cylinder turning round and round, inside which the still-tender threads were wound.

There was another new discovery in that same year, 1902: the German Müller discovered that it was possible to harden the threads and wind them at the same time, instead of splitting the operation into two stages. Müller's quick and economic process contributed greatly to the development of the viscose industry.

Although they had abandoned the quest for cellulose acetate Cross and Bevan took it up again when cellulose xanthate appeared to be just as difficult to produce. In 1894 the two researchers finally achieved a reasonable result, which they had patented, but cellulose acetate thread only became commercially viable when the American Miles discovered, also accidentally, that the technique of saponification in caustic soda, used in the cellulose xanthate process, obviated the need for dangerous chloroform to dissolve the cellulose acetate.

A more rapid process

Until 1985 the classic methods of viscose manufacture consisted of dissolving cellulose in sulphuric acid, then washing the gel thus obtained to isolate the pure cellulose fibre. A new process developed that year consisted of using another solvent which would be vaporized in a sterilizer (then recovered), shortening the delay in production by 80–90 per cent and making a considerable saving. So a remarkable revival in the viscose industry is expected by the end of the century.

Vitamins

Eijkmann, 1897; Funk, 1911

The next-to-nothing essentials

The nature and effects of these
organic compounds are well known.
Perhaps certain virtues are yet to be
discovered.

The beneficial effects of cod liver oil on rickets had been known since the 18th century, but the reason for this — that is, the richness of this product in vitamin D — was unknown. The first of the 20 vitamins known in 1986 — the word itself only appeared in the English language in 1912 — was guessed rather than discovered by the Dutchman Eijkmann in 1897. Eijkmann was working at the time in the Dutch East Indies (now Indonesia), where a disease called beri-beri, characterized by neuritis and polyneuritis (both conditions affecting the nerves) was rampant. As these symptoms only appeared in Europe in cases of acute alcoholism, which was certainly not the case in the sufferers from beri-beri, Eijkmann looked into another cause, which he identified as being the consumption of polished rice, the staple food of the population. Eijkmann managed in an experiment to reproduce the disease in hens, simply by subjecting them to an exclusive diet of polished rice. It was in 1911 that the Polish researcher Funk, after verifying that a diet of whole rice did not give rise to beri-beri, analysed straw and found in it a substance which cured the disease; this was thiamin, later called vitamin B1.

So it was Eijkmann and Funk who discovered the first vitamin. They were not, however, the discoverers of all the vitamins which have each been the subject of one or several specific finds.

Scurvy and rickets

The explanation of beri-beri led many doctors and researchers to suppose that certain unexplained diseases might also be caused by deficiencies. In 1912, therefore, the Germans Holst and Fröhlich submitted guinea pigs to an exclusive diet of oats and hay; this caused a deficiency disease comparable to scurvy, which was rife in earlier times among sailors and other people deprived of fruit and vegetables. Actually Holst and Fröhlich were simply endorsing observations which were already established fact and which, since the 18th century, had led captains to stock up on lemons and green vegetables. The value of Holst and Fröhlich's contribution was in determining which plants did or did not contain the necessary substance: ascorbic

---

Food and vitamin therapy

Most of the recommended daily doses of vitamins for human beings are largely covered by a balanced diet. These doses are, moreover, the subject of debate and are sometimes held to be higher than our actual needs. The international medical community points out that specific vitamin therapy, with the exception of vitamin C treatment, may expose people to serious disorders of hypervitaminosis ('overdose') and must therefore only be undertaken on medical advice and under supervision.

acid, also called vitamin C. (The term 'vitamin' had been invented the previous year by Funk, on the basis that the necessary substances were amino acids.) Vitamin C was not to be isolated until 1932 by the British scientists King and Waugh.

In 1913 the Americans McCollum and Davis and the Britons Osborne and Mendel independently discovered a third vitamin, a lack of which led to disorders of vision and of the mucous membranes in particular; this was vitamin A, present in animal fats, fish oils and the carotenoid pigments of certain plants, like the carrot. Its chemical nature was not to be discovered for another 20 years.

Rickets, another deficiency syndrome, was the next to stimulate research into a 'missing' vitamin, which the British researcher Mellanby discovered in 1918 thanks to animal experiments and which the American McCollum rediscovered in 1922. However, this discovery was essentially completed by the Germans Steenbock, Hess and Weinstock, who revealed in 1924 that the synthesis of this vitamin, vitamin D, is carried out through the skin under the effect of ultraviolet light.

The year of the discovery of vitamin D, the Britons Evans and Bishop, working on oily substances, in particular vegetable oils such as that of wheatgerm, discovered in them a vitamin, named E or tocopherol; this was isolated by the Britons, Evans and Emerson, in 1936 and chemically identified in 1938 by the Germans Karrer, Salomon and Fritzsche. Deficiency of this vitamin is almost unknown.

In the 1930s the interest of the researchers began to go beyond the field of deficiency diseases, although experimentation continued to bear essentially on variations in diet. This was how in 1934, having subjected hens to a vegetable-free diet, the Danish researcher Dam demonstrated that this deficiency brought about spontaneous haemorrhages. Following this an anti-haemorrhagic substance, a naphtoquinone, which Dam called vitamin K1, (Koagulations-vitamin) was isolated in the green leaves of vegetables. Paradoxically a substance encouraging haemorrhages was to be found in fermented clover; this was dicoumarol, exploited in the 1950s as a rat poison.

B2, B3, B5, B6

In 1933 the Germans Kühn, György and Wagner-Jauregg found a substance necessary for the health of the skin and the eyes: riboflavin, or vitamin B2.

The discovery, in 1927, that pellagra is caused by a deficiency obviously led to research into a dietary cause. Pellagra is a serious disease characterized at first by dizzy turns, headaches, a burning in the throat, diarrhoea, tiredness and unusual sensitivity to light, followed by nervous and mental disorders, possibly including paralysis, delirium and suicidal tendencies. It is particularly rife in the populations of Central America. The vitamin, deficiencies of which cause the condition, has been found; it is nicotinic acid, also called niacin or vitamin B3. The cause of pellagra can be compared with that of beri-beri, as it arises where diets are based essentially on maize. Vitamin B3 was discovered in 1937 by the British researchers Madden, Strong and Woolley and Elvehjem.

In 1936 the American Birch and the German György found that certain erosive deteriorations in the skin of rats was linked to a dietary cause. Two years later five laboratories had separately isolated the preventative factor: it was vitamin B6, known in the three forms of pyridoxal, pyridoxamine and pyridoxine. This vitamin improves conditions which are not essentially due to deficiencies.

In 1933 the American Williams discovered a vitamin which was later shown to be essential to the balance of certain enzymates; lack of it leads to whitening of the hair, skin disorders, and characteristic burning sensations on the soles of the feet. It is a component of the co-enzyme A and is panthotenic acid or vitamin B5.

An incomplete discovery

In 1927 the German Boas had demonstrated a curious fact: the addition of raw egg white to a balanced diet led to disorders in animals. It was subsequently established that this phenomenon was due to the presence in the egg white of a substance called avidin, which blocked the action of a vitamin, biotin. Biotin had long been known since it is essential in the culture of bacteria and yeasts, but which was only established as being necessary to humans in 1950. Biotin is also known as vitamin H. In the 1930s, while research into the question of vitamins was at its height, a common test

---

Vitamin A and cancer

Numerous studies, notably by the Americans and Japanese, lead us to think that vitamin A derivatives may play a preventative role against cancer, and in particular cancers of the colon and the lung. A very far-reaching programme of research on this point was undertaken in 1986 by the American government, under the aegis of the National Cancer Institute. An excess of vitamin A can be toxic and this vitamin does not lend itself to self-medication.

---

for determining whether or not a recently discovered substance was suitable for classification as a vitamin consisted of testing it on bacteria cultures. The British researcher Day, in 1938, and his compatriots Hogan, Parrott, Snel and Peterson in 1940, had discovered what seemed to be a new vitamin: it was necessary for the growth of bacteria and it prevented anaemia in monkeys and hens; this was folic acid or pteroylglutamic acid, also called vitamin B9. Its nitric acid nature was established in 1946 by the Lederlé Laboratories. It is in fact a distinctive vitamin in that it is synthesized by the bacteria of the digestive tract.

In 1927, the Britons Minot and Murphy had discovered that pernicious anaemia could be effectively treated by a diet of raw liver or by injections of extracts from this organ, but it was not until 1948 that the active element was extracted in crystalline form: this was cobalamin or vitamin B12.

Also classed among the vitamins are substances with a vitamin-type function, such as the bioflavonoids, (also called vitamin P) and the unsaturated fat acids, which were sometimes known as vitamin F.

Discoveries in the area of vitamins remain to be made, as is evidenced by those concerning the anti-cancerous action of vitamin A; the antioxidizing effect of vitamins C and E, and the effect of vitamin C on the adrenal glands; all of these were accomplished between 1971 and 1982 and are the subject of further study.

Vulcanization

Goodyear, 1839

When rubber stopped melting

*The artificial hardening of rubber
ushered in a whole new industrial
era.*

The introduction of rubber (see p. 189)
to Western markets sparked off an interest
which lasted for nearly a century. Not only
could latex be used to remove pencil marks,
as was first noted by the Briton Priestley,
discoverer of oxygen, but it also waterproofed clothes, shoes and luggage. In 1842
alone, the United States had imported
nearly half a million pairs of rubber 'shoes'
manufactured in South America, by the
Indians, who poured latex on to wooden
moulds provided by the American
importers.

However, since the 1830s interest in
rubber had begun to decline. In fact, latex
became sticky in summer, and hard and
crumbly in winter. Attempts were made to
make it more stable by coating it with a
solution of rubber and turpentine, but this
was not very effective.

Industrialists were reluctant to renounce
latex and researchers begun to look for a
stabilizing treatment. They tested several
substances suitable both for adhering to the
latex and for stabilizing its surface and,
in 1832, the German Ludersdorff and the
American Hayward found, independently,
that sulphur reduced the tendency of latex
to become sticky. However, sulphurized
rubber still did not display the desirable
stability which it was hoped would make it
akin to leather.

In 1830 the small-time American indus-
trialist Charles Goodyear also concerned
himself with this problem and for a time
thought that he was well on his way. By
treating rubber with *aqua fortis* or nitric acid
he achieved a surface hardening which was,
however, temporary In 1836 he won a con-
tract from the American government for
the manufacture of mail bags but, shortly
afterwards, it was realized that, in high
temperatures, the latex treated with *aqua
fortis* again exhibited its unfortunate tendency to become viscous.

A happy blunder

Goodyear would not be beaten and, in
1837, working with Hayward, he bought
the latter's process from him, although it
had been shown to be rather unsatisfactory.
For two years he tried to perfect the sul-
phurization of latex, but in vain. In January
1839, his clumsiness came to his aid: he
upset on a stove a container into which he
had put latex, sulphur and, as an experiment, white lead. When the mixture
cooled, Goodyear could see that it had
acquired the much-sought-after stability
without a compensatory loss of elasticity.

Goodyear's secret was simple: it consisted
of polymerizing the latex with sulphur and
white lead by the action of heat. The
annoying thing was that the sulphur powder
which formed on the surface gave the
game away, and the process was rapidly
imitated. Goodyear obtained his American
patent in 1844 but was beaten to it in
Britain by Thomas Hancock who, having
analysed products made according to
Goodyear's process, reached an understanding of its principle and took out a
patent himself in 1843. It had indeed been
his friend Brockedon who coined the word

vulcanization after observing Hancock's experiments with sulphur, sulphur being of volcanic origin and Vulcan, the blacksmith god, supposedly inhabiting volcanoes. Goodyear fought until 1852 to have his prior claim to the invention recognized. He ought to have made a fortune; instead he was imprisoned for debt in Paris in 1855 and died a ruined man in New York five years later. The vulcanization factory which he had founded in Paris had collapsed.

However, the success of vulcanization was stunning. A whole industry was created in the United States and the Western world: tramcar bumpers, raincoats, balls and toys, boots, and tyres from 1845, the year that Thomson invented the pneumatic tyre. The present range of products in vulcanized rubber has obviously grown tremendously, including padding for seats, covering for various pipes, vehicle shock absorbers, rubber tubing. ...

Improvements

It would be wrong to see Hancock merely as an imitator of Goodyear. He did, in fact, improve the rubber industry considerably by the invention of the masticator, which grinds the raw rubber to give it more pliability, and of calenders and presses as well as moulds which have, since 1846, facilitated the manufacture of waterproofed materials with a fine layer of vulcanized rubber and objects moulded under pressure, good enough for industrial uses.

In 1906 the American Oenslager introduced the use of organic accelerators, which allow for a rapid vulcanization with more formless products. Other chemical products increasing the resistance of vulcanized rubber were discovered in 1923 by the Americans Winkelmann and Gray, and in 1924 by their compatriot Cadwell.

Astonishing physical characteristics

We can measure how invaluable the discovery of vulcanization has been by comparing the physical characteristics of pure latex and vulcanized rubber. The degree of elasticity for vulcanized rubber is between 275 and 350 kg per square cm, against 20–40 for pure latex. Hardness on the tensiometer goes from 20–30 to 40–45 respectively. The capacity for displacement without distortion is reduced, though, since it falls from 800–1200 per cent to 675–850 per cent. The distortion resulting from a displacement of 200 per cent during 24 hours falls from 75–125 per cent to 3–5 per cent.

Furthermore, vulcanized rubber is practically impermeable to water and gases, and its constituents are not affected by oxygen, bases and acids, or organic solvents. Vulcanized natural rubber exhibits the best resistance to cracks, crazing and erosion.

Ebonite

In 1851, increasing to 50 per cent the amount of sulphur necessary for vulcanization, Goodyear discovered ebonite or hard vulcanized rubber, which had considerable success, particularly as an electrical insulator, until the advent of plastics.

X-rays

Hertz, 1892; Röntgen, 1895

A piercing light

The discovery of high energy photons
preceded by only a year that of
radioactivity, of which they form a
part.

Few discoveries have so clearly marked the transition from one era to another as did that of X-rays. Before it physicists believed in the existence of a 'luminiferous ether', and thought that the hydrogen atom represented the smallest mass that could exist. Several observations and discoveries preceded that of X-rays, however, and history has thus only assigned it a place comparable to that of one star within a constellation; it represents, none the less, the discovery of the fundamental structures of matter.

How X-rays are produced

Atoms bombarded by electrons can have their inner electrons liberated by collisions with the bombarding electrons. This leaves a gap which is filled by an electron which is further away from the nucleus. However, this electron has too much energy and must lose some in order to fill the vacancy. It does this by emitting an X-ray photon.

New tubes

One of the favourite instruments of physicists in the second half of the 19th century was the discharge tube, a closed glass bulb inside which a cathode (negative electrode) emitted an electric current (in fact a beam of electrons) towards an anode (positive electrode). In 1858 the German Plücker discovered that an area of the tube became phosphorescent and that this area could be moved by a magnet. The movement could be explained by an electromagnetic effect, but the phosphorescence remained a mystery. Twenty years later the Englishman Crookes improved the tube and compared the bundle of cathode rays to a current of molecules 'in flight'. The electron then being unknown, Crookes did not take this idea any further.

The first discoverer of X-rays was, without doubt, the German Hertz, already famous for his discovery of the waves which bear his name (see p. 176). In 1892, in fact, Hertz discovered that cathodic rays could cross thin metallic sheets, an observation repeated two years later, the year of Hertz's death, by the Frenchman Lenard, with the aid of a tube of his own design, including a window of thin aluminium.

Passing through matter

Almost everything was set for a definition of X-rays. Only one factor was missing: the capacity of X-rays to pass through, from a distance, a given thickness of opaque matter (a capacity noted by Lenard, but in such insignificant proportions that it did not incline the scholar to pursue the observation). This property was discovered by Röntgen by accident during an experiment.

Röntgen repeated the experiments of his predecessors, but in detailed and methodical fashion. It was already known that a screen coated with platinum cyanide of barium had become phosphorescent, exactly like Plücker's tube, when it was placed very closely in front of a Lenard tube. However, the experiment had never been tried with a Hittorf-Crookes tube, a type of tube with thick glass walls. Röntgen began by testing whether the cathode rays would pass through the glass or not. In order to check this he surrounded the tube with black cardboard, made the laboratory dark, and then passed a high voltage electric charge through it. Röntgen got ready to switch off the current in order to bring the screen smeared with platinum cyanide closer to the tube. The screen was nearly a metre from the tube and not right beside it, as might be supposed necessary for it to become phosphorescent. It was then that, to his great surprise, Röntgen noticed in the darkness a mysterious luminosity, whose intensity corresponded to the discharges across the tube. Röntgen struck a match and discovered that the luminosity was coming from the barium platino-cyanide screen, located outside of the cardboard casing.

The unknown ray

Lenard had suggested, however, that the cathode rays scattered in a thin layer of air; thus this was a new phenomenon as the cathode rays could not have travelled that far. Röntgen set about describing this phenomenon with great excitement. He noted that all the substances were more or less transparent to cathode rays, but some, like barium platino-cyanide, the calcium compounds, and the crystals and salts of uranium, continued to emit light after being exposed to the mysterious rays, which Röntgen, for this reason, called X-rays. These rays, which made impressions on photographic emulsions, were not very susceptible to refraction and reflection and could not therefore be concentrated by lenses; furthermore, contrary to Plücker's observation, they were not deviated by magnetic fields. One very important fact is that the X-rays discharged elements electrified by ionization of the surrounding air. This last observation was to allow the British scientist Thomson to find a method

for measuring the intensity of X-rays the following year.

Finally, Röntgen noted that the emission of X-rays on to a metallic body triggered a second emission, the intensity of which varied according to the atomic mass of the body. This is why platinum produced more X-rays than aluminium. 'Paper is very transparent.' noted Röntgen. 'Behind a bound book of about a thousand pages, I saw the fluorescent screen light up brilliantly...' On 28 December 1895, Röntgen sent a paper on X-rays to the Society of Physical Medicine at Würzburg. In a few weeks he was famous.

It was medicine which benefited from the first application of X-rays. It was becoming possible to study the skeleton and the position of foreign bodies in an organism, but the real impact of Röntgen's discovery took several years to become apparent, even after Becquerel discovered radioactivity a year later in 1896 (see p. 133). At first it was thought that X-rays were identical to cathode rays, which was wrong. (Cathode rays produced inside the tube are accelerated electrons, whilst X-rays are high-energy photons produced by the electrons.) When, in 1906, the Englishmen Barkla and Sadler, measuring the coefficient of diffraction of the carbon in X-rays, found that the carbon has six electrons, their find strongly encouraged a belief that X-rays do in fact consist of electrons, as in Thomson's theory. Achieving the diffraction of X-rays, this time using a crystal, the Germans von Laue, Friedrich and Knipping, together with the Englishman Bragg, managed to verify its electromagnetic nature. However, also in 1906, Einstein explained that the photo-

---

Electricity, chemistry and mineralogy

Arising within the space of a few years, from 1895 to 1903, the discoveries of X-rays, radioactivity and radium are a good example of the way in which research carried out in different fields can then come together. The discovery of X-rays in fact derives from research on electricity, that of radioactivity from research in chemistry, and that of radium and polonium from both chemistry and mineralogy. All three come together with the emission of rapid electrons.

electric effect which produced X-rays, amongst other effects, was due to the liberation of particles of energy, photons, by the electrons. This was a revolutionary idea which took a long time to establish itself.

In 1925 the Englishmen Compton and Doan established the range of the wavelengths of X-rays: this spans $0.1Å–300$ Å and extends, in the short waves, into the spectrum of the gamma rays of radioactive substances.

Röntgen and the atom

Radium had been discovered more than a quarter of a century before and it was noted soon after that it emitted gamma rays, like other radioactive bodies (see p. 134); radium had also been used for many years in radiography. Crookes' tube and radium do have something in common: they both emit the same type of radiation, made up of electrons, and by accelerating cathode rays to a certain degree, rays analogous to those of radium are obtained. In a way, Röntgen discovered and produced artificial radioactivity 40 years before its official birth.

However, if the unitary nature of electromagnetic radiations has indeed been defined (Maxwell and Lorentz), Röntgen's pioneering role has strangely escaped the wisdom of observers ...

At that time, the chief concern was to establish the structure and mechanisms of the atom. Gradually, in the 1920s and 1930s, the following concept took shape, at the cost of abandoning several previous theories: it was that the atomic nucleus is surrounded by several layers of electrons, those nearer the centre having lower energy.

To study genes

Several American physicists completed an original and costly project in 1986 (£200 000 were spent) for the study of the structure of genes. It consisted of directing very powerful beams of X-rays produced by a particle accelerator or synchrotron towards different genes. The beams should allow for the identification of the specific structures of genes in a few days, instead of several months.

Index

Note: Entries in **bold type** refer to articles headed with the name shown.

INDEX